The Muse of Coding

This book gives students and experienced programmers a way to see coding as an art and themselves as artists whose personal views, experiences, and ways of thinking can make their programs better for themselves and their users.

This book shows in a good-humored and sympathetic way how the artistic and practical sides of programming are the same, delving into the methods of coding, the history of art, and the ways in which artists and audiences interact and benefit each other.

Not confined to a single language or style of coding, this book provides a widely applicable framework for people to learn what languages and styles work best for them at present and as the field evolves. It can be used as a classroom text or for personal study and enrichment.

Richard Garfinkle is a computer programmer and author of science and math popularizations and science fiction and fantasy novels. He is married to a visual artist. Richard attended the University of Chicago and majored in mathematics. He has been programming since the era of punch cards and paper tapes.

The Muse of Coding
Computer Programming as Art

Richard Garfinkle

CRC Press
Taylor & Francis Group
Boca Raton London New York

CRC Press is an imprint of the
Taylor & Francis Group, an **informa** business

A CHAPMAN & HALL BOOK

Cover Image: Alessandra Kelley

First edition published 2024
by CRC Press
2385 NW Executive Center Drive, Suite 320, Boca Raton FL 33431

and by CRC Press
4 Park Square, Milton Park, Abingdon, Oxon, OX14 4RN

CRC Press is an imprint of Taylor & Francis Group, LLC

© 2024 Richard Garfinkle

ISBN: 978-1-032-60607-1 (hbk)
ISBN: 978-1-032-60606-4 (pbk)
ISBN: 978-1-003-45986-6 (ebk)

DOI: 10.1201/9781003459866

Typeset in Palatino
by SPi Technologies India Pvt Ltd (Straive)

This book is dedicated to my late aunt, Barbara Feigenbaum, who brooked no nonsense and delighted in the diverse paths the later generations in her family took.

Contents

Section II Too Much Information

Section III   Foreground: Artist's Life

Acknowledgments

This book owes its existence to my wife, Alessandra Kelley. The two of us come from very different backgrounds and practice very different arts. We have spent nearly four decades getting closer to each other, learning to understand each other's arts and perspectives, and coming to see how the human world is both practical and artistic and that both of those views are one and the same.

This book also owes its existence to our children, who grew up in a house of mathematics, visual art, coding, writing, and music. Each of them developed their own arts, and we are always excited and interested to see what they make and what we can learn from them.

I also owe thanks to my brother David, with whom I co-authored one science and one math popularization. Working together to present the insides and outsides of math and science was invaluable in developing this text.

Lastly, I would like to thank my publishers and editors for making this book possible.

Section I

Everything Humans Do Is Art

1

All Art Everywhere since Time Began

Humans make things and humans use what humans make. These facts underly both art and engineering. Humans explore, humans experiment, and humans reveal the results of exploration and experimentation. These facts underly both art and science. Humans abstract and humans model reality with their abstractions. These facts underly both art and mathematics.

In recent years, art and STEM (Science, Technology, Engineering, and Mathematics) have been viewed as in opposition, as if there were some conflicts between the illuminatingly beautiful and the illuminatingly practical. For some people, this supposed opposition deepens so far that they assert that the concepts of beauty and practicality are themselves in conflict so that the more beautiful a thing is the less useful it is and vice versa.

The origins of this illusory conflict can and should be studied. But that's not what this book is about. This text is about erasing this false dichotomy and introducing the practitioners of one of the most recent fields of art, computer programming, to the awareness that they are practicing an art. The goal of this book is to help programmers see their field and their personal work as an artistic endeavor because this view will help them do better work and more enjoy the doing of it.

To begin, we will, as all wise artists do, take a step back to gain some perspectives on art, humanity, and practicality.

Art History

As of 2023 CE, the oldest identifiable tools that archeologists have found are more than three million years old. These tools were stone scrapers and cutting tools made by our prehuman ancestors. The oldest objects that people are willing to call visual artworks are around 50,000 years old. The oldest known musical instrument is about the same age. The oldest evidence of humans wearing clothes is from around 180,000 years ago.

But these labels, tools and artworks, can mislead. They imply that art comes millions of years after tools. But what were our prehuman ancestors doing with those tools? What were they making? And what were they doing with those things that they made? And what kind of clothes were our ancestors wearing 180,000 years ago?

These questions and phrasings carry within them the implicit separation of practicality (tools and clothing) and art (painting, sculpture, and music). But does this separation, this contrast, this asserted dichotomy actually mean anything or is it just labeling?

There is a tendency in modern thinking to regard the utility of things as distinct from and in opposition to their beauty. Utility is artistically depicted as stern, hard, and earthy, while beauty is seen as airy and abstract, nebulous, and unimportant.

But if one asks whether clothes should feel good to wear, food taste good to eat, teachings be illuminating and memorable, houses be restful to live in, tools fit well in hands,

DOI: 10.1201/9781003459866-2

and so on, we find no clear boundary between beauty and utility. Far from being opposed, people want beauty in their utility and utility in their beauty. Indeed, for humans, the two concepts are so inherently entangled as to only be separable in theory, never in practice.

This can also be seen by starting from the other side. If we look closely at those aspects of human life that most people would explicitly identify as arts, drawing, painting, sculpture, music, song, and dance for example, we find that each of these has its own practical benefits.

Images and sculptures can aid memory and can present people, places, and things that are distant in a way that makes them imaginable. Music and dance give rhythm to actions, making it easier for people to work in coordination.

And then there are stories. Stories can bring the past to the present and can open up possible ways to consider the future. They can carry minds to distant places and show ways of action as yet unconsidered.

And, of course, all of these can be enjoyed, and people's lives are made better when they enjoy what they are doing. Enjoyment gives energy to humans in action. Enjoyment can bring the mind's attention to what it is doing. It can spur inventiveness and a willingness to return again and again to new variations of old possibilities. And enjoyment can be shared, creating social spaces that people like returning to.

Beautiful and useful are in a real sense different measurements of the same human actions.

Were this an art history book, we could follow the complex braided pathways of artmaking from those earliest artifacts through the diversity of human cultures until we reached the present day. Then we might dabble in predictions for the future so that future art historians could have a good laugh about how in error we were.

But this book has a different goal. We are trying to illuminate computer programming as an art. So, we are going to chart a more abstract course that looks at aspects of arts that are germane to that purpose. We will focus not on particular periods and practices but on art methods that have evolved into many art forms. We will see how computer programming fits in with them and how their perspectives and practices can help illuminate and improve the practices of programming.

Art as Illuminating Abstraction

The real world is a complex, information-dense space. Human minds operate in this environment by abstracting what is around them and working upon the abstractions.

This process of taking experience in from the outside world and formulating a mental object by combining what is perceived with what the perceiver knows and understands, this personalized act of making sense by abstraction and composition is the most basic artistic action. All art begins and ends with the interplay between reality and human mind that we call abstraction.

We will discuss this in depth in Section II, where we will make the process of abstraction more concrete and practical.

For now, let's look at the benefits and drawbacks of abstraction and how they lead to art as humans understand and practice it.

Abstraction allows the mind to focus in on certain aspects of the piece of reality being abstracted. It does so by only considering those aspects and ignoring all others, or at least trying to ignore all other aspects. This, as we will examine later, is not as easy as it sounds.

FIGURE 1.1
Elephant photograph.

Abstraction produces a simplified and emphasized object of mind. Each art has methods that help the artist determine what aspects to abstract as well as how to simplify the aspects and how to emphasize them. Different arts abstract different aspects in different ways because each art emphasizes different aspects of reality and so abstracts in ways are useful for its particular emphasis.

Abstracted mental objects can illuminate real-world objects and any other mental objects that the perceiver connects with the abstracted objects.

Here is a photograph of an elephant (Figure 1.1).

This picture already ignores nearly everything about elephants as a species: their environments, life cycles, evolutionary forebears, etc. It also ignores most of this elephant as an individual being, giving us nothing about its life and experience, how it ended up where it is, what it smells and sounds like, how it perceives and thinks, and so on.

What the picture does is present for human eyes an image of a large creature with a few emphasized features (including trunk and tusks) made using the art and technology of digital photography.

Here are a bunch of images of statues, drawings, and paintings that depict elephants using those arts, each with their own emphasis (Figure 1.2).

The above paragraphs in which elephants are discussed are themselves an abstraction of elephants into the art of writing. Each of these artworks reveals and emphasizes different aspects of elephantness.

All of the writing above presumes that you who read this know what an elephant is (even if you've never seen one in real life) and likely have various associations with the term elephant from whatever other elephant-containing art you've perceived.

If by some bizarre circumstance this book happens to survive to a time where the reader has no idea of what an elephant is, this whole section will be a bit more confusing and any readers from that time will not know what to make of this example.

Abstraction as seen from the artist's perspective is a matter of making something accessible to human perception and contemplation. It makes it possible for the human mind to consider, play with, enjoy, and learn from what is presented.

But abstraction is also a vital tool in science and mathematics, a tool used to make other tools.

FIGURE 1.2
Collage of pictures of elephant art.

Mathematics, Abstraction, and Tool Making

Mathematical abstraction involves creating numbers, shapes, spaces, and functions that correspond in a useful fashion to real-world properties, objects, extents, and processes. For an extremely elaborated discussion of this, see the author's cowritten work *X Marks The Spot*.[1]

The purposes of mathematical abstraction are twofold: to explore the implications of the abstract object and to make tools that will carry those implications into the reality that the objects are abstracted from.

Math is much more conscious of the act of abstraction than most arts are. From a mathematical perspective, we know that if we are looking only at the mass of an object, we are ignoring all the internal structure and shape of it, as well as other properties like charge and whether it's alive or not and how awkward sudden mass-to-mass interactions would be on a social scale.

Abstracting objects as mass alone gives us many implications to study when coupled with theories of force and gravity.

From these implications, we can and have made tools to measure and study the universe and tools to alter the world around us. Mathematical abstraction lets us put handles on the universe, pick it up, turn it around, shape it, and put it back down with new configurations.

Mathematically, abstraction is done to make it easier to make use of and understand what is being abstracted. Artistically, abstraction is done to make things easier to perceive, display, and understand.

These two abstractions, artistic and mathematical, are not separate processes. They are elaborations of the human mind's tendency to abstract in order to make it easier to perceive, make use of, display, and understand.

Refining abstraction so that it does all four of these is the province of a process that, in English at least, has a name with both artistic and mathematical meaning: Modeling.

Modeling and Representation

Modeling is the process of assembling multiple abstractions from some aspects of reality in such a way as to create a representative construct that reveals those aspects of reality and aspects closely connected to them.

This is usually done by using hypothesized relationships between those aspects, taking measurements of those aspects, deducing other aspects using those relationships, and depicting them in an illuminating fashion. Therefore, both of the illustrations below depict acts of modeling (Figures 1.3 and 1.4).

A model relies on four things to be usable.

1. Accurate formulations of relationships between quantities.
2. The ability to accurately measure some of the quantities discussed.
3. The ability to use the formulations and the measurements to determine other aspects.
4. The ability to accurately reveal the results of step 3 to people who need the results.

You may notice that there are some very fuzzy terms in the above summary like relationship, formulation, accuracy, ability, determine, and reveal.

These concepts are deliberately broad. Their breadth and fuzziness let us generate models for a vast space of possible things in an evolving space of arts for an ocean of purposes that has always revealed itself to be deeper than we thought we were modeling.

So, let's focus in on the art this book is about and look at programming relative to this general description of modeling.

Programming is very good at step 3, using a combination of formulations (functions) and measurements (data) to produce new quantities.

FIGURE 1.3
Measurement of object and drawing from those measurements.

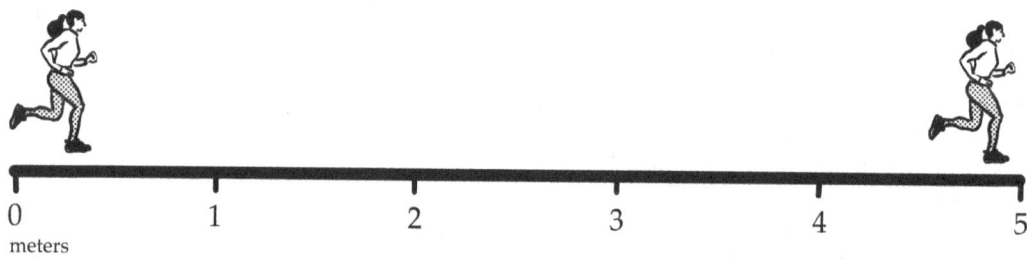

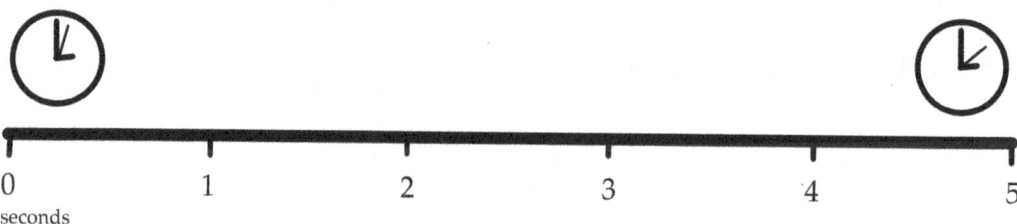

$$\text{speed} = \text{distance}/\text{time} = 5\ \text{meters}/5\ \text{seconds} = 1\ \text{meter}/\text{second}$$

FIGURE 1.4
Measurement of physical quantities for motion, list equations of motion, plug values into equations, and produce calculated quantities for motion.

It needs other arts and sciences to provide the relationships from which functions are coded (step 1), other tools to provide the measured data that will be processed (step 2), and specialized devices to display the results in human-comprehensible fashions (step 4). There will be more about all of that throughout this book.

There's one more little term that we let slide by, a term that covers messy and complex stuff: Reality.

There are vast numbers of books of people trying to define reality. We're not going to deal with that. We're working on the assumption that there is a real world that we are part of. We also assume that our observations of the real world have some correlation with what's actually going on out there. If you're interested in a more elaborated discussion of this that still does not try to define reality, see my coauthored book *Three Steps to the Universe*.[2]

It is important to note that a model need not be accurate to reality. A model only has to be internally consistent and capable of taking in measured input and creating revelatory output to qualify. This allows one to model unreal worlds (see all SFF, most video games, surrealist art, and many discarded theories of reality). It also allows us to deliberately inaccurately model the real world (see any map that depicts a spherical planet on a flat surface).

There are two further steps that we can add to the definition of the model. They aren't as necessary as the first four steps are, but they are vital both to the scientific method and to representational art.

Step 5: Check the results of steps 3 and 4 against what you are modeling, that is, see if what you produced does in fact look like/sound like/move like/feel like/taste like what it's supposed to be like.

Step 6: Change the model if it doesn't fit step 5 to the standards you need it to. Those standards might be the accuracy of measurement or clarity of illumination for the audience you are showing the model to or the ability to use the model in order to advance some mental or physical work you are doing.

Reality Augmented by Art

Let's look closely at step 4 in modeling. "The ability to accurately reveal the results of step 3 to people who need the results."

What happens to someone who receives an artistic revelation? What happens to the world they are perceiving once they have been given the results of a model of aspects of that world?

This question is easy to answer because you are experiencing it right now. You are looking at a book or a display device or listening to a recording or touching braille writing and your mind is generating meaning and understanding from what you are perceiving. But you can only do that because you have learned the language you are reading this in and are projecting meaning onto the writing or sounds.

Humans do this all the time. We live in an augmented reality in which we mingle our knowledge with what we perceive and create a view of the world that is humanized by perceiving objects and processes relative to our human lives.

Human minds evolved to generate this personalized AR environment. It makes the world around us meaningful, useful, and, in various ways, beautiful. Our sapience is the kind of sapience that generates augmented reality.

But the process as evolved generates many false positives. We give meaning to the not meaningful and adjust our perceptions so that what we perceive fits the models we are using to augment our perceptions.

We can hear random sounds and think they are human speech. We can look at patterns of craters or rock shapes and think they are human faces. These false positives are one of the things that make art possible. Because we can project recognition onto things, it is possible to deliberately create things that people will project recognition on to. Here are a variety of things that humans will recognize as human (Figure 1.5).

Human perception and memory use a wide variety of clues and complex neurological processing methods (that at the moment are only beginning to be understood) to generate the augmented realities we each live in.

We won't be looking into neurology itself, interesting though that subject is, but we will be looking in depth at many of the practicalities of this process and how understanding it can help with all artistic endeavors both from the perspective of artists and audiences.

FIGURE 1.5
Collage of humans drawn in a wide variety of styles.

Composition and Contrast

Art relies on our abilities to recognize and project recognition on what we are perceiving. Before there can be recognition, there must be some aggregate of perception information that a person can recognize. As we will discuss in Chapter 7, we live in an information-dense world. We are constantly flooded with the possible-to-recognize.

How in an ever-evolving chaotic welter of possible information do we find what we're perceiving? And how do artists use their art to create easily recognizable bundles of information?

Our minds put things together out of information and separate out what information we are focusing on from the rest of what's happening. This leads to two principles that all arts rely on:

Composition: Putting together.

Contrast: Separating out by.

Each art has its own techniques of composition and contrast. A large part of learning an art is developing one's own ability to employ these two.

Composition and contrast are not really separable/contrasting concepts. They are a paired awareness. When we perceive things, we both put them together and take them apart. Both synthesis and analysis are always happening in our minds to craft our awareness. As we will discuss in depth later, the ability to perceive an art is also an art, the art of audiencing.

There are several forms of composition and contrast that are worth analyzing separately because they recur and are basic to many arts.

We'll start with something that is both composition and contrast:

Apposition, which is putting things next to each other.

Because human memory and perception are associative, we will tend to compare and contrast things placed next to each other. This is the most basic way in which computer and human thought are qualitatively different. Two things near each other in a human mind affect each other. Two things adjacent in computer memory have no effect on each other. We'll be dealing with this in depth in Chapters 8 and 9.

Common forms of composition include:

Shared Properties: A group of things seen as sharing a common property are often treated as the same so the mind can easily compose them into a commonality. This one has to be used carefully because it is common to treat sharing one property as sharing many. For example, raspberries are red, tasty, and nutritious. Cadmium red paint is red and, therefore, a mind might think that it should be tasty and nutritious. This is also a common composition in prejudices. This form of composition shows how the false positive character of human perception can lead one to major errors, but is still useful for putting together a plate of food or a painted forest.

Blending: Blending composes materials in such a way that they produce a new material. It is a vital process in cooking, paint mixing, pottery, metallurgy, creation of colors on a screen, and coining of words.

Structuring: Structuring is the process of connecting objects in such a way as to make a new object that differs qualitatively from the components that make it up. It is related to blending in that it can make new materials (e.g., hydrogen and oxygen structured to make water). It is also used to make clothes out of fabrics and buildings out of wood, stone, glass, etc. Structuring by its nature creates and relies on an infrastructure/underlying connection that may be hidden from direct perception. Structuring is the most common composition method used in programming. We'll be looking at this in depth in Chapter 5, and referring to it pretty much throughout the entire book.

Metaphor: Metaphor is something of an extension of shared characteristics that deliberately draws upon aspects of one thing to illuminate another thing. Lips red as roses, lips red as flame, and lips red as cherries. All of these are metaphors that import different associations with them. Metaphors work best when they associate not only surface characteristics but infrastructural/hidden characteristics as well. The kenning "whale road" used in Beowulf to refer to the ocean carries the idea of the ocean as a means of travel. The kenning "widow maker" for the ocean carries the idea of its dangers. Both metaphors illuminate the hidden aspects of the ocean that a prospective mariner should be aware of. The ability to switch from one metaphor to another or to blend them together (Hamlet's "To take arms against a sea of troubles") is invaluable in illuminating multiple aspects of what one is talking about.

Persistence: This one is a little subtle. Human awareness composes in time. We watch things move and change and form connections between them. We associate the thing we saw ten seconds ago with the object we are seeing now and think of it as the same object even if it's moving farther away from us. Persistence is one of the phenomena out of which we build our ideas of time. We will delve more deeply into persistence and its relationship to programming in Chapter 2, the next chapter. Note that the idea of a next chapter, or a next anything, arises from our concepts of persistence.

Common forms of contrast:

Perceivable difference: This is the most basic. If things register differently in our senses, then we can distinguish them. Different tastes, colors, notes, volume levels, textures, temperatures, etc. all create contrasts that can be employed to create art. But our senses are only so good. Our ideas of differences are often learned or varied by previous exposure (i.e., persistence).

The idea of the color blue covers a broad and fuzzy region of colors and different people will classify something in the aqua region as blue or green. Whether something feels cold or warm depends on what temperatures we are currently acclimated to and to our inner bodily processes. There's a common experiment with three containers of water – one hot, one cold, and one lukewarm. How the warm one feels to a hand depends on whether that hand was in the cold or hot container before going into the warm one.

Perceivable differences can be elaborated and tuned by cultural attitudes and personal experiences. We'll examine this in a broader context later in this chapter.

Separation in space: This is one of the most obvious. Two things that are at different positions will look distinct to us. Visual art relies as much on spatial separation as it does on spatial apposition. This has a few tricky elements in that differences in position can produce the illusion of difference (or sameness) in size.

The fact that the sun and the moon have similar apparent sizes to the Earth is an artifact of their actual size and actual distances. That artifact led to millennia of direct association between them, creating an artistic pairing that does not correspond to the distinct realities of star and lump of rock. On the other hand, both objects form a neat pairing to create time measurers, the sun to measure days and parts of days and the moon to measure months and parts of months. Their apparent commonality of size and shininess made it easier for humans to make tools out of their apparent orbiting. A lot of utility can be created by putting together things that only look like they fit together.

Separation in time: This one is more complex than separation in space. Our experience of time is not symmetric. We can't go back to a time we've been at before. We accumulate experience in one direction. And though our memories are not organized chronologically, the experiences that generated them happen in the local chronology we call personal experience. So having A before B is not the same as having B before A.

Making art that relies on separation in time (such as writing a book or composing music or teaching a class) requires an awareness of what is likely to happen if someone experiences B before A versus A before B. There can be a temptation to overload or underload parts of the experience, to go too much into detail when a thing is introduced. But often it is wiser to introduce an idea, then let it sit and be mentally digested, and close with a suggestion that more will be dealt with later in Chapters 2 and 3.

Separation in environment: The circumstances under which we experience things have radically different effects on how much we enjoy them and what we get out of them. Experiencing something for the first time in an enjoyable context can create associations of that thing with enjoyability. Experiencing it again under miserable circumstances can help remove some of the misery or replace the enjoyable associations. Environment and context are high-density forms of apposition. The tailoring of an environment and its experience is an aspect of art common to architecture, clothing, furniture, music, dining, and well, pretty much everything.

Separation in environment is also, along with separations in space and time, a vital consideration in the making of art. How, when, and where will you work upon what you are making? This will affect the quality of your work and the inspirations and distresses you will feel while working. This too we will discuss a great deal more in many places to follow.

Separation in status: This one is largely social. Objects, actions, and people are accorded statuses in societies, cultures, and subcultures. How someone has been taught to think about those statuses can affect how they perceive the things associated with that status. People internalize these early on and are often surprised when not everyone shares their separations. They are often also surprised that their own views evolve over time.

Separation in meaning: This is a process of distinguishing what underlying or overarching understanding is being depicted in the art. There are terms and ideas that cover qualitatively different subjects. The word "nature," for example, might mean physics, or inherent qualities, or inborn characteristics, or romantic woodlands or violent storms. Distinguishing which nature one is referring to requires clarity and illumination of art. A technical field will usually develop a technical vocabulary to make such distinctions; but outside the field, a fair amount of work is often needful to make clear what one is talking about.

Distinction in properties: This is when objects have differences in properties that are not necessarily perceivable nor inherently meaningful. The actual as opposed to the perceived sizes of the sun and moon are distinctions in properties. That one is a star and shines by fusion and the other is a big rock that shines by reflected light is both a distinction in properties and meaning.

Distinctions in properties are relative to what one is doing with the objects under examination. The biological differences between fruits, vegetables, and herbs are distinctions in properties. The culinary differences are distinctions in different properties with the same names.

Art Is Personal and Cultural but Never Universal

While the creation and enjoyment/utilization of art are shared characteristics of all of humanity, and the underlying methods of that creation are also shared, individual works of art are always local to their makers and their time, place, and culture of origin. They are also individual in who likes them and how they are liked.

De gustibus non est disputandum is a Latin maxim that roughly translates as there is no arguing about taste.

There's also the French saying *chacun a son gout*: Each to their own taste.

There are similar expressions in many other languages. These ideas are usually said as a way of ending the discussion of taste or ironically starting arguments wherein people object to the sayings and try to claim that there are objectively good and bad tastes.

We're not going to do either. Instead, we're going to examine what this personal quality of art means on practical, personal, and cultural levels.

The neurology of taste is being studied and examined now and as usual we're not going to dig into it. We're also not going to go too far into the psychology.

Our approach will instead be to illuminate what the late Sir Terry Pratchett, author of the Discworld novels, called a "substition," a thing that is true that hardly anyone believes is true. The individuality of taste is true, but most people believe that there is right taste and wrong taste and that those who have wrong taste are either in need of education or are somehow mentally or morally wrong.

There is a practical reason why this basic fact is so often disbelieved. When most people see someone liking what they like or disliking what they dislike, a composition is created in the mind of the observer. They likely feel a fellowship and understanding of that person. They feel a mental proximity, a drawing closer to them and perhaps a space of possibilities to share with them.

If, on the other hand, people see someone else enjoying something those people could never enjoy or not enjoying something they do enjoy, then a contrast is created that creates a mental distance, a separation between them.

But why would the composition create art in the observer's mind, but the contrast not create art? Why is there a problem here?

The difficulty lies in what to do next after observing the other person's liking or disliking. What memory does the observer form from the fact of the contrast.

As I laid it out above, one form creates closeness and another creates alienation because the contrast is treated as an uncrossable chasm of tastes. But by adding one step to the process of awareness, one further augmentation to reality and the chasm can be bridged.

Composition: I like grapes. They like grapes. I can get and serve grapes to both of us.

Contrast: I like grapes. They like pineapples. Added step: In a situation where I would eat grapes, I should serve them pineapples. I will get and serve both grapes and pineapples.

The new step can be abstracted into the general principle: What is A for me is B for them. Therefore, in circumstances where I would like A, they would like B.

This means remembering particular tastes and interests of the other people one is interacting with. Because of the ways humans form memories of each other, this can be highly efficient. For most people, the better they know other people the clearer and more efficient their memories of those people become.

This approach works well on an individual case-by-case basis when the person one is dealing with knows what their own tastes are. But how do people determine what they personally like and dislike? This is generally done through a process of experimentation, trying things, and seeing how one feels about them.

All experimental methods require a test for outcomes. Generally, things done that are to one's tastes feel good in the doing and leave one feeling happier and generally more energized. Things that are against one's tastes are unpleasant in the doing and leave one feeling less happy and either drained or numb.

Unfortunately, people can be bad about experimental protocols, and other people can interfere with results and recording of results. Furthermore, as we will deal with in detail later, human memory is not great for data recording. There are a number of things that can interfere with the process and cause one to be bad about performing the experiments and remembering the results.

Here are a few such problems that can interfere with the memories one forms of one's direct experiences. There are many others, but these should give a sense of the space of them.

One more bite: Any enjoyable activity will eventually become too much. If the person testing something does not pay attention to when they've reached their limits, they might keep going and enter the unenjoyable zone. They might then keep going even further, trying to recover the initial enjoyable feeling, and end up with a memory of the experience being overall unenjoyable.

Shame: An action that was enjoyable might be something that is treated as socially unacceptable and the person doing it might therefore associate shame with the action and not realize that they actually liked it.

Should dos: The opposite effect of shame, doing something unenjoyable that one is told one should do and should enjoy the doing. This can produce false memories as well, wherein the person overlays a claim of enjoyment over the fact of non-enjoyment.

The above experimentation applies not only to the art one is trying as an audience member but also to what art one is trying to make. One can try to make too much of one kind of art (e.g., writing too many stories with the same plot) or be ashamed that one is doing a socially unapproved art, or feel that one's skills should be better spent on something socially favored, etc.

Shifting from the individual taste to a larger-scale social perspective on what arts to practice and which to audience, a number of questions emerge. What arts are available for someone to experiment with? What art does an individual have access to and when in their lives do they have access to it? What can they try and when can they try it?

The answers depend on when and where people grow up and live their lives, what art is available within their culture, and what would they need to do to have opportunities to sample it? This is automatically entangled with the question of what art/art making will the person inevitably be exposed to because of the time, place, and culture of their upbringing. This can be complicated by what art/art making is being deliberately taught and how is it being taught. A person can think that they do not like/are no good at doing a particular art if the teaching methods do not work for how they learn.

Children are exposed to a huge amount of art in their first years. The languages spoken around them and the clothing, furnishings, and architecture of their homes and larger environments are ambient to their existence. The nearby terrain and how it is described and traversed are part of their journeys and everyday routines.

There are also the music, songs, and dances they hear and take part in, and of course, there is the food they are fed on. Childhood is being the audience to a vast panoply of arts.

As for arts a child learns to practice, they will likely learn how to speak, read, and write. Depending on culture they may learn to sing or play music, to draw, to play particular games and/or sports, to watch videos, and to see views of distant and nearby places. They will play with particular toys and learn to use specific hardware and software. They may be told stories of their families, peoples, perhaps religion or religions, and so on. They may make friends and learn to converse and inquire and come to see other people and points of view. They will be taught practices meant to guide them toward adulthood and given ideas of what their grown-up life might be like.

They may, depending on where they live and who is nearby, have a monocultural or multicultural childhood. They may be part of a subculture or minority within a majority culture. They may have little or much access to publicly available art in the form of parks and museums, theaters, and so on. They may learn one or more than one language. They may live in one or many places, may travel, or never go far from their place of origin. All of these shape their individual experiences.

Childhood is art dense and experimentation happens at high speeds and at all times. No wonder children get exhausted.

Children learn very rapidly, so rapidly that they often don't notice that they are learning anything at all. They absorb culture and formalize it and their relationship to it in their minds without realizing that anything strange or amazing is happening.

As people grow older, they obtain more ability to determine which art they will interact with and start shaping their own personal experimentations. But all of that will be done within the context of one or more cultures they are dwelling in or visiting. Their personal experience of art and art making will always be entangled with the environments of those arts and the ability to access them.

Substitionally, people absorb a great deal of art and art making and formulate complicated views of all of that without much sense of the processes they have been through or the results they recorded. Their likes and dislikes are not simple.

Normalization, Fashions, and Aesthetic Theories

One aspect of human thinking is the ability to acclimate to current environmental and social conditions. Acclimation is useful to humans because it lets us go from environment to environment or experience the changes of seasons with relative ease. It's a form of relativity, wherein we implicitly reset a center point for current conditions.

Acclimation cannot make one enjoy what one does not enjoy, but one can get used to ongoing unhappiness and treat it as what to expect. At the other extreme, it is possible to get so used to what one enjoys that having any enjoyable thing taken away can feel like deprivation.

Therein lies the difficulty, because acclimatization does not make a value judgment on the different environments one is shifting through. It is easy to go from the inherent process of acclimatization to the mental error of normalization.

Normalization is the assertion that what one is used to is how things should be. Normalize suffering and one may work to remove joy from one's life. Normalize having what one wants, and one may treat not having it as if something had been stolen from one regardless of why one does not have it anymore.

Normalization takes a vital mental process of relativization to changing environments and treats it as if one were in an unchanging environment that has always been and will, or at least should, always be the same.

This is a paradox of the mathematics of relativity. Every frame of reference has an origin that can be seen as moving through the universe or the universe can be seen as moving around it. Each frame of reference defines its position as the origin of a universe-delineating coordinate system.

If one loses the awareness of relativity or if one was never taught it in the first place, it is easy to normalize one's awareness into a completely self-centered perspective that instead of picturing the universe asserts that the universe is the picture one makes, that the rotation around one's life is absolute, not relative.

There are a large number of individual, social, and political problems caused by this kind of normalization. Those aren't directly what we're focusing on at the moment, although they all matter in the history and evolution of art, and we will come back to some of them in later chapters.

What we are going to look at are two phenomena in the history of art that arise from the normalization of bundles of artworks and schools of art: Fashions and theories of aesthetics.

Fashions are perspectives on certain kinds and groups of artworks and styles that a culture or subculture will assert to be the arts and styles that everyone (or everyone who is anyone) should enjoy. Fashions usually flow in waves arising from one place or subculture and propagating outward through overlapping groups of people.

Rarely are fashions the result of one artist's works. More often, a group of artists (often with no direct connection to each other) are seen as all being artists of the same fashion. Fashions mutate as they change, and different artists take them up.

While fashions are attributed to artists, they are more often created by audiences. The fashionable works are picked up by some people and then jump to others due to shared excitement or peer pressure or both. There are attempts by publishers, movie studios, clothing designers, and others to try to push fashions, but these don't work anywhere near as well as many people think.

Fashions flow in cycles. They move and propagate in waves. While they are going on, they are high energy normalization, giving a sense of excitement to the fashionable and distress to those who want to be fashionable but cannot be.

Fashion cycles have been documented in various ways. Clothing fashions of past centuries have been documented in paintings and sculpture, although a lot of the painters and sculptors did not know enough about clothing-making to understand what they were depicting (Figure 1.6).

In the 20th and 21st centuries CE, movies and videos did a good job accidentally recording the fashions in clothing, hairstyles, body types, and makeup of the times in which they

FIGURE 1.6
Clothing and painting instances.

were made because the normalization of fashion meant that nearly every such video was full of artifacts of its time.

Which actors were chosen to portray which characters and what costumes, makeup, hairstyles, etc. were employed in the videos were always of the time and place of making. This was especially visible in movies that attempted to be historical. People who knew the histories could see clearly that the portrayal was not accurate to the times shown.

In the same eras, fashions in cooking were documented in cookbooks. A student of culinary arts could easily track the emergence of recipes, plating styles, and so on using such books.

Fashions in books were documented in bestseller lists. Fashions in music were documented in top hit lists for radios and later downloads.

It might not be obvious that all of these fashions are artifacts of normalization. But one thing that is common among fashions is that years later many people who followed them look back on what they were doing and wonder why they wore those clothes, ate those foods, and had that hair and makeup style.

It is also true that the most popular works in any art of a time period are rarely the ones prized in later eras. Only a few pop culture sensations (e.g., Shakespeare) become the exemplars and teaching resources of later generations of artists. If one looks at any of the bestseller lists from a generation or two back, one will find few works that are still popular and many that have faded from memory completely.

Fashions rise up and pass away in waves. They can make it harder for artists to make their work uniquely, and they can push artists and audiences to take up art that does not really make them happy.

Aesthetic theories arise from the same error of relativization that cause fashions, but they can cause much bigger problems and extend for centuries or even millennia, in contrast to the life cycle of a fashion which is usually on the order of one to ten years.

Nearly all aesthetic theories are based on the premise that there is a single universal concept of beauty or a hierarchy of concepts of beauty with a true idea of beauty on top and all others beneath it and inferior.

Some aesthetic theories were imposed from outside by academics. Some grew from the inside of arts when a particular technique or style was asserted to be universally applicable and not to be questioned. And some arose from a concerted effort of cultural one-upmanship that insisted that one people's art was inherently superior to that of all other people because those people were inherently better/smarter/more beautiful, etc.

The aesthetic theories that cause problems are those that ignore the idea of personal taste and assert that there is an objective beauty that is in some way an aspect of objective truth. This error is embodied in the last lines of John Keats' poem written in 1819, *Ode on a Grecian Urn.*

When old age shall this generation waste,
Thou shalt remain, in midst of other woe
Than ours, a friend to man, to whom thou say'st,
'Beauty is truth, truth beauty,—that is all
Ye know on earth, and all ye need to know.'

To be fair to Keats, the poem itself is an embodiment of inspiration – one artist making art in response to another artist's work. And poetry is not meant to be taken as academic teaching. But to also be fair to Keats, he clearly thought what he was saying was true.

The aesthetic theory that grew among the literati in the UK during the 19th century with Oscar Wilde as its most voluble exponent actually thought this equation of truth and beauty was both true and beautiful and pushed the idea that certain works of art should be experienced by all of humanity as perfect illuminations of objective ecstatic beauty.

But an aesthetic theory doesn't have to be that dramatic to cause problems. There have been odd geometric theories that asserted that certain shapes and ratios were inherently beautiful and that everyone should draw, paint, sculpt, and make furniture, clothing, and architecture using those ratios. The golden ratio has been fashionable on and off for this (Figure 1.7).

There have also been and still are theories about the meaning and usage of color. Carl Jung, the early psychologist, had ideas that particular colors meant particular things and favorite colors told a lot about the mind that favored them. This can interfere with visual artists' abilities to use the colors they need for what they are depicting.

Aesthetic theories have worked at times to suppress growth, change, and experimentation in various arts. The attitudes of the French Academe toward the Impressionists near the end of the 19th century created a good deal of misery for many artists who are now seen as brilliant and whose work is far more prized than anything the Academe of the time approved of.

Aesthetic theories of some cultures have also been used to label the art of different cultures 'primitive.' This act deliberately dismisses the skills, care, and meaning that goes into making any and all art.

Aesthetic theories also crept into STEM fields, even reaching as far as mathematics wherein the earliest fractals mathematicians created were deemed pathological because they didn't look pretty against the backdrop of Euclidean geometry.

To understand why none of these universalizing theories can work, we need to dig deeper into the fundamental relationship that makes artwork at all.

FIGURE 1.7
The golden ratio.

Artist and Audience

If everything humans do is art (as my wife, this book's illustrator, Alessandra Kelley is wont to say), what does it mean to be an artist?

If human minds always perceive a reality augmented by art, what does it mean to be audience member?

If these questions are asked like this, the answer to both is, "to be human" which is dramatic, mystical, and floofy, and impressively unhelpful.

But if we narrow the focus of these questions and ask,

What does it mean to be an artist making this specific artwork? and

What does it mean to be audience for this specific artwork?

we can start to get our hands dirty and our senses focused and examine what, in practical terms, artist and audience mean and what is needed to be competent at artisting and audiencing.

Arts, Responsibilities, and Playspaces of the Artist

Each human mind is unique, but all humans share similar needs and biological and neurological structures. Between this unique each of us and this shared all of us, there lie a diversity of possible artworks that can be made by an individual or group of individuals that can be uniquely experienced by an ever-expanding space of other humans.

The basic responsibility of an artist is to make art that arises from their uniqueness in a form that can be shared and benefited from by a diversity of other unique humans.

There is an odd paradox in this responsibility. Each unique mind has unique perceptions and augmentations of reality that are obvious to it, but that others may not see at all. Fundamentally, art that expresses that which is obvious to one and not obvious to others is some of the most valuable art.

Art that arises from an unshared perspective that can be shared is most likely to inspire others because it expands the space of ways others can see and act in the world. In effect, what is obvious to one person can create possibilities for other people that would not otherwise exist in their lives and minds.

This is true whether the artist is working alone or with others. An actor playing a part in a play or movie or video is one of many artists working to make the art. They will need to do their work cooperatively with all the other artists on stage and off. But they will bring their unique skills and perspective to the character they are playing and the interactions that character has with the other characters and the lines they are saying. Whatever the art, however many the artists, there is always room for uniqueness and, if the art is properly made, the opportunity to bring forth that uniqueness.

In order to do this well, an artist needs to learn these materials and techniques of their art that work for that artist. The artist must practice using and experimenting with those materials and techniques to expand their capabilities in order to evolve their own awareness of and expression of what is obvious to them.

This learning and experimenting is an ongoing process that never ends, which has good and bad points.

The good points are that we are always growing in what we can do. The major bad point is that many artists are unhappy with their work and often can't stand to look at what they did years (or even days) ago.

But while competence and uniqueness can make art that looks good to an audience, they do not suffice if the artist is ignorant of what their art is depicting. This is quite visible in one of the most famous artworks in the world, the Sistine Chapel Ceiling painted by Michelangelo.

The male presenting figures have stylistic similarities but are pretty anatomically accurate. The female presenting figures aren't. Michelangelo had biases in what anatomy he studied and was very biased in what he was interested in showing. Those biases are evident in his work.

The ignorance and knowledge of an artist is always visible to anyone who knows well what is being depicted. Ignorance jumps out at the part of the audience that knows what the artist is ignorant of. Unfortunately, ignorance artistically depicted can also imprint itself as the illusion of knowledge on the audience that does not know. Ignorance and prejudice are easy to propagate.

For example, during the early decades of the 21st century, there were a number of TV shows about police work that featured forensic science and implied a level of accuracy for its methods that was not in any way true. The actual fields that were being woven into the shows were and are highly specialized so few people in the audience knew that the shows were dramatically fictionalizing this area of police work.

Whatever your art is showing, you should try to know to at least one level deeper than you are showing. It's always good to know the underlying anatomy before you depict a figure.

That stricture can sound a bit intimidating if one is starting out to learn an art. But the thing about learning is that there's a lot to learn and a lot of mistakes to be made to learn from. And the thing about art is that one is always learning. Improving competence, learning from others, practicing, and studying away one's ignorance is an ongoing process.

It's important to get better, but it's also important to find practices of improvement and sources to learn from. It helps a great deal if one has practices and sources that one enjoys.

Finding these can be a tricky process of trial and error. There are a lot of sources out there such as schools, classes, teachers, books, videos, and online references. Finding what works for you can take a long time and a lot of trial and error because learning is a matter of both self-discovery and exploration of what is out there. And while you will likely find what works for you now, as you grow and change and your art grows and changes, you will likely later learn that what worked at one time no longer works later. That's human life for you.

Once you find the methods that work for you now, you'll likely find that you enjoy learning, practicing those new methods and getting better at old ones, finding new insights, exploring new and well-traveled spaces, and making practice works can be a great deal of fun. The deeper in the art you go, the more your mind is likely to bubble up ideas and possible things to play with until the whole world becomes part of your mental playground.

When that happens, you will find that you are always making. A mind that has integrated an art into itself sees the world relative to that art and starts considering how to depict things or make tools for things or turn things upside down and inside out. Even if you rarely produce specific artworks for other people, your mind will be constantly involved in the world-transforming process of making art.

This can be a beautiful and sometimes ecstatic experience. It can also be a tricky thing to deal with. People practice different arts and each person's view of their art is unique. Each art has its own jargon, and so, one can find oneself not speaking the same language as other people in one's life.

This process of separation that comes with learning an art has the risk of making the artist into an artiste.

Everything humans do is art; therefore, every human is an artist. An artiste is someone who thinks that being an artist makes them special.

Special is not the same as unique. Each of us is unique. Special is a part of a false dichotomy between common and special. The falsehood is the concept that there are common people and special people.

It's easy to create this error by judging humanity relative to a single characteristic, such as *do they practice the same art I do?*

This question immediately divides humanity into two groups: Those who do and those who don't. All that is needful to create the error is to slap the label special on those who do and common on those who don't.

But, of course, this question can be asked for any art, and therefore this division can be created many times. Each person in humanity can be labeled common for each art they don't practice and special for each art they do.

All that is needed to avoid the error and therefore not develop an erroneous sense of superiority is to not slap the labels of common and special on the sets of people who don't and do practice the art. Without the labels, it becomes pretty simple to relate to people practicing other arts.

> Her relationship to playing baseball is like my relationship to coding. She's going through a difficulty with her batting like the one I went through last year when I didn't think I was getting better at interface design.

And then onward to shared human interaction across an apparent divide.

Arts, Responsibilities, and Playspaces of the Audience

All human art is made relative to human awareness and perceptions.

This may seem obvious, but there is a tendency to think that an artwork is inherently perceivable and understandable as that art, as if the artist's intent was magically imbued in the work and, therefore, birds and beasts and stones and clouds and gravity itself would perceive and comprehend the meaning therein.

That's not how it works. Even if we remove the drama of stones and clouds and gravity and just look at animals, we quickly learn that the arts that affect us do not affect other animals in the same way. Many birds see things faster than humans do, so a bird looking at a video made for humans sees a series of still frames.

For a more extreme example, I recall a dolphin pool in an aquarium. The pool had painted scenes of underwater life. Those scenes would not fool the dolphins, who perceive by echolocating sonar. Those walls sound like walls, not like open seas.

Okay, so fine, is art universally experienceable by humans? Is intent present in the art itself so that all humans will get the manifest intent?

Clearly not. If one does not know a language one cannot understand a conversation in that language or read stories in it. If one does not like a particular food, it does not matter how well a cook prepared it.

There are deeper examples in the field of archeology where artifacts are unearthed and if their usage is not well known already (like pots and knives), the discoverers shrug their shoulders or sometimes classify them as ritual implements as a vague catchall.

Even things we are good at recognizing (such as human faces) are not universally readable. In European art before the 20th century, smiles that showed teeth were a sign of moral depravity or low social class, or racist. In the early 20th century, this changed to the modern idea of smiles showing happiness. By the middle of the century, photographers were encouraging smiles so that pictures would imply happiness.

What does this tell us about audiencing?

To understand and benefit from art, the audience must be skilled in a variety of receptive arts. People need to learn to listen and read and taste and make sense of forms and shapes. They also need to learn to use tools in order to appreciate the quality and utility of those tools.

There are a wide variety of audience skills, most of which are not given names. Reading is one of the few that is named, largely because reading was for a long time an art practiced by only a small set of people in those societies that had reading and writing. Reading was a specialist professional skill until the idea of widespread literacy caused vast sea changes in cultures.

Unfortunately, this means that few of the skills needed to audience are taught anywhere and people often find it difficult to determine what they need to be able to do in order to appreciate and enjoy the arts they might enjoy.

Different kinds of music need different ways of listening. Different cuisines are eaten and tasted differently. Different visual arts need to be looked at with different perspectives and expectations.

The important thing here is to remember that audiencing takes its own skills, and that one may need to learn the basics before one can find out whether one enjoys a thing or not.

However, there is no need or benefit in someone learning arts that just do not work for that person. Someone who gets motion sick will not be able to learn skills to enjoy riding roller coasters. A person who dislikes seafood will not be able to appreciate bouillabaisse no matter how well made.

Along with receptive skills, learning the contexts of artworks is vital to understanding them. Humans have a tendency to assume that other people's lives are like their own unless they learn differently. Different cultures and different times produce a wide variety of human experiences.

If one is experiencing art from a different place, time, or culture than one's own, it is beneficial to learn about that context to reduce the risk of projecting presumptions outside the field of one's own experience. See the matter of smiling above and multiply that by the vastness of human diversity to get a sense of how easy it is to misinterpret every aspect of what one is experiencing.

The more one explores the arts one enjoys, the more the world fills up with them. The world can be full of images and stories and music and gardens and sports and games and whatever other arts one enjoys.

A wise Muppet once said, "Listening is the first step and the last step." He was talking about learning to be a musician, but the principle applies to every receptive art. Senses must be opened and tuned to the art one will be experiencing in order for the world to fill up with that art. If one does not know how to listen for the beauty within the sound, one cannot hear the music. If one does not attune one's perceptions to see how various peoples have depicted the world and humanity, one cannot see the diversity of human-perceived shapes and spaces. If one does not learn the ways that various people have told stories and taught lessons, one cannot become expansively aware of how humans have shown human awareness of human thoughts and actions.

But just as artists can become artistes by seeing the arts they make as special, audience can become aesthetes by seeing the art they appreciate as special. The error and the structures are the same, as is the solution. Knowing what one likes and why one likes it is good, but it does not set one apart from anyone else. There is no common or special audiencing.

Every human audiences uniquely. Helping others find what they like rather than trying to get them to like what one likes is good audiencing. Trying to get everyone to try what one likes and disdaining those who don't is not good audiencing. Sitting alone with the certainty that one is the only true appreciator of the art one likes is a waste of art and life.

The Chaos between Artist and Audience

The relationship between artists and their audiences can be chaotic. By chaos, I mean the subject of chaos theory, that is, functions that can produce wildly different results from small changes in input parameters.

Chaos is pretty easy to generate in reality. It's happening in the weather right now. Even seemingly controlled phenomena like induction between currents in wires can lead to chaos (Figure 1.8).

Small variations in artworks can produce wildly diverse reactions in members of an audience. This is because human minds and memories are associative and complex and contain the learned experience of an ongoing lifetime within the framework of a unique way of thinking and one or more cultures that evolved by the actions and artworks of a wide swathe of people.

Even if we narrow all that down to one person looking at one artwork, that same person can have very diverse reactions to the same artwork in different contexts or times in their lives. A story can be formative for a child and when looked back upon in adulthood, it can seem obvious or distressingly incorrect. One's first exposure to a piece of music can be

FIGURE 1.8
Chaotic magnetic field between two wires.

strongly influenced by the emotions running through one's mind at the time one is hearing it. And one bad experience with a food can turn one off it for life.

But it isn't just audience's reaction that is subject to this chaos. It is also created on the artist's side. Small changes in attitude, mood, and perspective can affect how artists express their ideas and place them into the composition and contrasts of their artworks. A new insight can reshape a work completely without anyone in the audience knowing that any such thing happened behind the scenes.

The back and forth of the art and the reaction to the art create the intertextuality/the social context in which the art is perceived and reacted to.

This chaos is most visible in the art form that involves the highest density of artists and audience members: Conversation is the art in which all participants are both artist and audience.

Tracking the flow of conversation can be fascinating. The play of ideas, subjects, illuminations, inspirations, collaborations, and frustrations vividly shows the chaos and makes clear its benefits.

The nonuniformity of expression and the reaction of speech, listening, and response shows how well art can spark and affect human thinking and action. It also shows the effects of changing who is involved in speaking and listening and how wide the space of responses and illuminations can be. Conversation is quick art in microcosm. It is the weather chaos of human interaction.

But what about other, slower arts? How does the chaos manifest in the evolving space of museums and libraries, public performances, and fashion waveforms?

For the longer scale chaos is more tectonic, more like the reshaping of continents and occasionally the arising of land in volcanoes and the drowning of it in tsunamis.

Art–audience interaction happens in many cycles and time scales. We mentioned this briefly in the discussion of fashion.

Arts can change from small sources like individual artists or audience members. The marriage of Catherine de Medici to Henri II of France in 1547 CE brought elements of Italian cuisine as it then was into France, shaping French cuisine for centuries after.

It can also flow through from cultural interaction. Rhyming poetry entered various European languages from Arabic sources in the Middle Ages in what is now Spain. Rhyme displaced alliteration to take its place with rhythm to form the dominant form of western poetry.

Art styles and works have also been looted at many periods in history and their display has inspired later artists in the nations that stole them. Forced migrations and slavery have also brought arts from one culture to another. Desperation and scrabbling in mud and mire have given birth to cuisines that eventually achieved worldwide fame and enjoyment. Art history is not nice and pretty.

In the various tectonic waves, from the smallest scale of conversation to the largest evolution over thousands of years, the distinction between artist and audience can vanish. Each human making art is audience to more art than anyone could make in myriads of lifetimes. All humans are audience to far more than they can make, but they also make far more art than they think.

Everything said, every action taken and witnessed, everything humans do can inspire and be of use to other people. This includes the reactions of others to our art, which can feedback into the later art we might make.

We are awash in the seas of art and are ourselves makers of those seas.

But at no point do we need to be lost in those seas.

Making Art Personal

The relationship between artists and audience, the chaos between them becomes both deeper and simpler if the artists and audience know each other, especially if they know each other well. All arts have inherent subgenres of art made for people one is close to. Love poetry, home cooking, toys made for closely related children, and games played between friends and family, all of these have added personal dimensions and entanglements that produce more complex chaos than art made for an audience one only theorizes exists.

People one knows well one is more likely to know the tastes of. Thus, it is easier to predict if the person or people one is making for will like the art. There is also a different feedback effect of the reaction. Art appreciated by someone one cares for can itself be more deeply appreciated.

There is also a deepened effect if the audience knows the artist well. Experiencing the works of someone one knows adds one's awareness of those people to the experience of the art. Ideas one has discussed with them or heard them speak about emerging in their art can expand the art and its meaning.

This process of expanded art meaning also applies to obtaining art that someone else made for someone one is close to. Gifts can be transformed because the obtainer of the art did so with the awareness of the person's tastes and interests.

Art made, art appreciated, art gathered, art shared.

Everything humans do is art.

Let's see how that applies to programming.

Personal Exercises

1. Consider what if any arts you approach as an artist and which arts you approach as audience. How as an artist do you think about your audience? How as audience do you think about the artists whose work you enjoy?

2. Consider ways of modeling things that you have found worked well for you, both artistic and scientific. Which ways of looking at reality have made things clear to you and illuminated possibilities that you wished to consider and play with? Examine which models have connected for you and consider what that reveals about how you think and approach the world around you.

3. This exercise depends on sight. Take two objects, place them next to each other, and look at them together. Watch what associations your mind makes between them.

Now separate them by about half a meter and see what creating space between them does to the association. Increase the distance enough that you need to turn your head to look from one to the other and see if the associations are still there or if they have been replaced. Then place the objects so one is in front of you and the other behind and see the separation that comes from needing to turn your whole body to perceive them. Return the objects to being next to each other, watch the return of association, and see if the associations are the same as they were before you did the separation.

4. This exercise depends on hearing. If you play a musical instrument, feel free to use that for this exercise. If not, use any piece of software that generates notes. Play two notes in rapid succession. Pay attention to how the notes come together in your mind. Play the same two notes with greater and greater time separation until they no longer come together in your perception. See how long a time that is.

5. Take one movie, book, or video that depicts something you know little about. Do a little research into that subject and see how much misinformation is depicted. Do the research using sources that are actual professionals in the field; do not rely on internet commentators.

Testable Exercises

1. Write a paragraph describing one tool you have that works well for you and what about it works for you. Then do the same about one tool that does not work well for you, but that you need to use for something you do.

2. Look up at least one tool from any era up to 200 years ago. Briefly research how it was used and which people used it. In your own words, write a brief description of the tool and its place in the society that used it.

3. Pick one kind of real-world object, plant, or animal (but not a human being) you are very familiar with. Find at least four depictions of that object in different art forms. Write a sentence or two about what aspects of the thing each artwork abstracts.

4. Pick one model that you like to make use of and describe how you use it and why it works for you. Use the four-step description of a model and lay out how the model approaches each of those steps and why your ways of thinking and working fit with that model.

5. Examine the rise and disappearance of one fashion you (or someone you are close to) were heavily involved in. Describe the fashion and its influences on other aspects of culture and how it was normalized.

6. Take one movie, video, or book that depicts something you know well. Discuss what elements were accurately depicted, what inaccurately, and whether that accuracy or inaccuracy mattered to the art/story.

Notes

1 X Marks the Spot: The Lost Inheritance of Mathematics. Richard Garfinkle and David Garfinkle. AK Peters/CRC Recreational Mathematics Series) 2021.

2 Three Steps to the Universe: From the Sun to Black Holes to the Mastery of Dark Matter. David Garfinkle and Richard Garfinkle. (University of Chicago Press) 2008.

2

What Programming Does Best

Each Art Does Something Better than All the Others

Writing reveals the insides of things.

Sculpture shows moments.

Pottery makes empty spaces.

Drawing shows edges.

Painting shows blending.

Cooking draws in reality.

Puppetry shows spirit.

Music evokes emotions.

Dance shows cooperation.

Clothing reveals motion.

Conversation generates ideas.

Each art can do all of these things and much more, but each art does one thing more easily and more simply than any other art.

When working in a given art, it often helps to center one's thinking around what the art does best and from that derive what actions to take in order to make the artwork one is making.

Audiencing a particular art can also be beneficial to focus on what the art does best in order to compose one's awareness, and from that see if one can enjoy this particular artwork.

So, what does programming do best?

Because we're examining this question in a written book, we'll be looking inside something to find what I think is the answer.

We could dig into my thought processes and construct an artificial path of reasoning that sounds like a plausible narration for how I came to this conclusion.

But the fact is that when I thought about this question, an answer popped up immediately and I had to go backward to figure out why it made sense. Once I did that, I contrived a forward-going way to elucidate the awareness. The elucidation is a more useful piece of writing than either the artificial path of reasoning or the backward examination.

DOI: 10.1201/9781003459866-3

All of that took a couple of paragraphs for you to read. Imagine how much pottery or how long and complex a dance I would have had to do to try to communicate which particular inside action I'm going to write about. That's the fundamental efficiency that comes from focusing on what an art does best.

The answer I saw begins by examining motion and its relationship to human perception.

In Motion and Moving

Our entire universe is in motion. Motion is inherent in our entire universe. Momentum and angular momentum are conserved quantities. They can neither be created nor destroyed, only redistributed. This is true from every frame of reference. Every perspective on the universe experiences motion on all scales from the fuzzy uncertainty of the picoscopic to the unfolding growth of the entirety of the universe.

Physics began in part as the study of motion, and our understanding of the universe has evolved with our understanding of motion. We exist in motion and in a very real sense we are made of motion.

But that doesn't mean that the world looks to our perceptions like a whizzing spinning welter of objects moving. Made of motion does not mean moving. It means that motion is fundamental to the infrastructure of how all things work.

Our concept of objects arises from perceiving things that are moving together as if their common movement made them one thing. The motions of the cells that make up our bodies, the molecules that comprise those cells, the atoms that make up the molecules, and the subatomic particles that make up the atoms are subsumed in our awareness of macroscopic objects as aggregates of shared motion.

This awareness creates the illusion that we perceive still objects. Our visual arts have for millennia seemed to depict still objects. But that is an illusion. The perception itself is in motion, and our concept of objects arises from an interesting mental phenomenon called persistence that relies on motion.

The neurological phenomenon of persistence and the related psychological concept of object permanence can be dived into if that interests you. But broadly speaking, the awareness of persistence evolved, and it is useful to us. Persistence lets us keep our eyes on things. It lets us put things down and pick them up later. Objects are an illusion, but they are a convenient illusion that our ways of thinking evolved to make use of.

Persistence is what programming does best.

Persistence and Programming

There is a practice to see how much our visual perception relies on motion. A sighted person sitting still and focusing on a single point in space for a period of time will generally find their vision turning fuzzy and greying out. Rather than getting a sharper vision of the objects around them, they will stop seeing them.

This happens because our vision relies on our eyes moving. Our sense of shape of things comes from taking in multiple images over time across bits of space and constructing a sense of the object from those images. Our awareness also takes in the fact of awareness in motion and the environment moving and evolving relative to us. We expect changes in what we perceive and take account of them when we construct our evolving awareness.

Mentally, we have the idea that we are looking at objects in motion. Our senses and mental constructs compose from the motion what we have perceived and anticipate the motion we expect to perceive. This applies to sight, hearing, touch, and in more subtle ways, to taste and smell. This combination of past and anticipated future motion also strongly influences our kinesthetic senses.

Our minds' ability to connect images perceived over time is why videos and animation work for us. A succession of sufficiently similar images presented in sufficiently rapid succession is blended in our minds into a continuous awareness of objects in motion.

This aggregation of multiple pieces of sensory data over time lets us catch hold of the concept of a persistent object. It lets us turn away from what we are looking at and expect an object to likely still be there when we turn back. We even expect things to be where we left them sometime ago.

Persistence is fundamental to our idea of objects. We rarely consider how much work our minds do to construct and maintain persistent objects from an environment of ongoing motion and change.

If we look at physical artworks in terms of persistence, we see that they take effort to make, maintain, and keep clean, and are very difficult to duplicate. They take effort to pass around from person to person (large artworks like the Pyramids of Giza are especially nuisancey to transport).

If we contrast these with virtual objects, we see that virtual objects take quite a lot of work to create. But they're very easy to store, replicate, display, and pass around (it would likely take you very little time to find an online image of the pyramids of Giza). Computers have revolutionized storage and replication. They have made it possible to put away and easily recover persistent things. And they do this better than any other art.

So, why is it easy for programmed objects to act like we think real-world objects should when they don't? Why is there virtual persistence in a world of ongoing change? How did we create an art that makes an unreal reality of objects in contrast to our real unreality of objects?

To answer this, we need to consider where persistence exists and what the storehouse of human persistence (i.e., human memory) is like, unlike its virtual analog.

Human Memory versus Computer Memory

Human memory is a complex subject that is currently being unraveled by neurologists and neuropsychologists. We're not going to approach it from their perspectives. Instead, we'll look at it from the convenient fact that the author and all the readers of this book are humans with human memory. So, we'll approach it from a shared experience and the examination of memory practices that have been developed over time.

To begin with, human memory is associative. It works by conjoining ideas and perceptions. Because of that, it is not separable from human thinking and not separable from human imagination. Our memory composes from what we perceive and what we already know and how our personal tastes and interests influence how we think about what we think about.

What we remember evolves in interesting ways. We evolve understanding and skill by practicing. We rarely remember the particulars of our practice. Do a lot of addition problems and one is likely to learn how to add, but unlikely to remember what specific additions one did. The problems themselves were consumed in the memory and the ability to add was built from that consumption.

This kind of learning, the learning of how to do things, can change our thinking so much that it becomes extremely difficult to not do what we know how to do. If you are reading this visually, take a moment and stare at this text and see if you can see the shapes of the letters without reading what the text says. Try to make the text disappear and see only the patterns and contrasts. Some people can do this. Most of those people are skilled visual artists. But for the rest of us, it's very hard.

Before we knew how to read those patterns and contrasts, what were we seeing? But few of us remember how texts looked before we could read.

If you are listening to this as an audiobook, you can try to hear the sounds as sounds but not as language, and that's even harder than not reading.

What we remember is continually in motion. New and old thoughts can bump into each other and create sudden realizations. A thing one read decades ago can blend into something one has just perceived and pick up stray thoughts from something one vaguely remembers someone saying. The ensuing bundle of concepts can lead to a whole new chain of thoughts that bubbles out into art.

The efficiency of human memory lies not in its accurate persistence, but in its miscibility and malleability.

Bluntly, we neither perceive nor recall what we have actually experienced. Instead, we are creating/making art of what we are experiencing, art that will hopefully be of use to us in living our lives, interacting with other people, and getting more from the art we perceive and make.

All of this is, of course, a simplification, and does not touch on matters of traumatic memory, false memory, muscle memory, the effects of disease and aging on memory, and so on.

Nor does it delve into various arts and methods humans have created and employed to make better use of their memory.

The chaotic artistic character of human memory is something that people have at times shied away from. A mythology of memory developed over the last couple of thousand years based on the extremely odd idea that humans perceive everything accurately and record things perfectly. Coupled with this is the blatantly false assertion that any person should be able to accurately recall all aspects of their experiences and lessons at will and on command.

Aristotle expounded an early form of this myth by saying that memory was like wax used to take impressions from what the person was experiencing. Yes, this is why we talk about the "impression" we get from things.

Another metaphor was that human memory is like writing. The experiences are inscribed in the mind. But writing was created as an aid to human memory, an ability to record what would last beyond a single mind. Writing, archives, and libraries became metaphors for human memory even though their function was and is to do what human memory doesn't.

This myth escalated over time, and was prevalent when the first computers were being designed and built. The concept of memory as an archive of accurate impressions that could then be accurately recalled was widespread, and it was this kind of thinking that guided the construction of digital memory.

In effect, computer memory was made to be an idealization of a kind of memory that humans did not have but told themselves and each other that they did have.

Computer memory is radically unlike human memory because people were working from inaccurate ideas of human memory. In a sense, computer memory was made to be better than human memory, which was seen as an imperfect form of memory. This is one of those things that happens when people have attachments to theories. They regard the theory as perfect and the instances that don't fit it as wrong.

The psychological and neurological studies of memory have become better because the fields are now investigating what's actually going on in our minds and brains rather than what people wanted to believe was going on.

We're not going to cover this evolving knowledge, but having placed the associations in this text, your mind will either take it in or disregard it (the fact that one of the exercises is about this may bias what you do).

If your mind takes in that the study of memory is evolving, it may be interested enough to go elsewhere to learn about it. You may already know more about it than I do and be correcting what I'm saying or be annoyed at me for errors I've made. You might also be annoyed at me for or amused at what looks like a digression but is also an illumination of the subject under discussion. Your reaction to what I've written here and the memories you form from it are individual acts of making art happening in your mind. You are creating memory from this text and your thinking, and I am making use of that to try to create a sideways, partially meta view of what you are thinking and doing with your thinking.

But this text, currently being written on a computer and stored on other computers, once published is not changing. It is stored according to the model of perfect impressions and recalled according to the model of perfect recall.

Where it is stored in computer memory is unimportant because it will not form associations with other files unless actively affected by software. The storage itself is compartmented.

This file will persist unchanged and recallable in a way that is impossible for human minds.

That unchanging character, that unblending, still quality of what lies within memory is the first reason programming is so good at persistence. Data in storage remains in storage until acted upon by software (or one of those messy physical things like breakage, but that's a hardware not a software problem).

And, herein, lies the irony. The fact that computer memory is unlike human memory is a feature, not a bug.

Jumping to a larger perspective, there are three kinds of tools: Tools that enhance human actions, tools that substitute for human actions, and tools that do things humans can't do at all.

Clothes and ladders and furniture and containers for carrying things enhance human actions. Ground vehicles substitute for human actions, as do drawing implements and cutting tools. Flying machines, cameras, sound-recording devices, and buildings do things humans cannot do.

Computer processing substitutes for human action (more about this later). Computer memory does something humans cannot do: Remember without change. Making something according to an erroneous model of human memory gave us tools that do what we cannot and that, to quote Robert Frost, has made all the difference.

The Ease of Copying and Varying

Copying physical art is at best extremely difficult. Most of the time it's impossible. The process of hand copying a visual artwork is an exercise in creating a new work of art. The copyist (or forger, if dishonest) is making their own art from their awareness of what they are seeing and using their own skills. The history of forgery is an interesting one because it shows the difficulty of copying and the ability of people to convince themselves that a bad copy is an original.

Copying texts by hand is equally difficult. People misread and reinterpret all the time. Studying copies of old manuscripts shows how much the human mind will always be playing with what it is seeing. Copying without changing is a mind-taxing skill.

But copying without changing is phenomenally easy if the copying thing has no idea what it is copying. If it is only copying bit by bit, it can copy and check the copying easily. Did that 1 become a 0? Fix it.

The casual ease of computer copying can be seen in the existence of the copy command that is so prevalent in so many programs. How much better this tool is compared to all earlier methods of duplication can be found in the two commands that currently accompany copy: cut and paste.

Cut and paste, in times past, meant to literally cut something out of a real physical, made of paper, document with actual physical scissors or a knife and then to physically paste the cut out using real glue onto another piece of paper. This was a slow, messy process, and it was annoying if you later realized you needed to move what you had just pasted to some other part of a document.

Computer cut, copy, paste, and all the other standard editing tools allow for controlled deliberate changes in virtual objects over time. It also allows one to cut a piece of document, and casually try it in different places to see where it works best.

Memoria versus Addressing

The ability to control the changes made in computer persistence versus the more chaotic artistic changes in human persistence relies in part on one of the deepest errors in the false model of human memory.

Because human memory is associative, we can recall things by what we associate them with. We use objects and circumstances associated with what we need to recall in order to help recall them. These are called aides memoire or memoria.

Statues, gravestones, portraits, and so on are used to remember people. Scents can be strong memoria; a brief smell of something familiar can recall aspects of one's past. The author Marcel Proust took this to extremes in his multivolume work, *A La Recherche Du Temps Perdu (Remembrance of Things Past)*, when he has his main character recall a vast stretch of time by dipping a madeline (a small sponge cake) into tea.

Memoria draw out associations from other memoria, but in human thinking what is drawn out is composed of other things. No human memory is really isolated. The words I am typing are memoria, but they are blending together in your mind to make a new meaning from the sentences I am typing rather than each word pulling up a separate set of associations.

Memoria can also go in multiple directions because association makes each memoria a thing rememberable and each memorable thing is also a memoria for anything associated with it. Seeing a person one knows, one might recall them at an earlier time wearing a particular outfit. One might also see the outfit on another person and recall the first person. One might see either the person or the outfit and recall an event where one saw the person wearing the outfit, then recall things that happened at the event, and so on in a vast cascade while the tea goes cold and the sponge cake soggy.

Computer memory takes the idea of memoria and flattens it into the simpler concept of addressing. Each location in computer memory has an address and a number that means that location. The address is used to store and retrieve the contents of that address.

The contents cannot be easily used to find the address. If one has the information and seeks the memory/address, the computer would need to search through addresses and check what is stored there in order to find where the thing is.

Addressing is a nuisance for humans to deal with and is only directly addressed in low-level computer languages. High-level languages conceal addressing behind the process of naming the data one is storing and retrieving. The computer assigns an address to the name, stores the data at the address, and when called upon finds the address for the name, and the data for the address and recalls that.

The names we give to chunks of data form a pivot between the human memory of the programmer and the computer memory of the computer.

If a program I'm writing is going to store a census of wombat population, I might make a variable named 'number_of_wombats.' If I'm feeling unnecessarily clever, I might give it a name of a wombat character in something I've seen or read. If I'm feeling overconfident about my ability to remember what I'm doing I might just call it 'w.'

The variable name has human memory associations for me and is assigned an address by the computer to store the information. Meaningful association is connected to a meaningless but precise location.

This connection is crucial to the way computers handle persistence.

Because we can pull or store an arbitrary amount of data out of a computer by using an address/reference, it is easy to retrieve what we are working on, modify it, and then either put it back where we got it from or put the modified version in another location.

We can create something and then modify it several times to create a stored evolution (Figure 2.1).

If we are to show these in rapid (rapid by human eye standards not by computer standards) succession, our persistence of vision will see them as one moving image. Ease of storage, ease of change, ease of recall, and no internal association give us the optimum tool to fool our human persistence into thinking that we are perceiving an ongoing transformation like the ones we see all the time in reality.

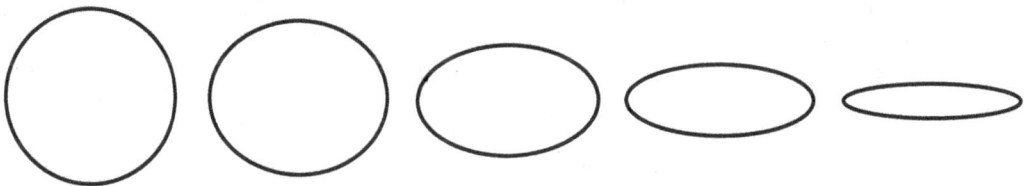

FIGURE 2.1
Five fairly simple computer-generated images, each slightly changed from the previous one.

Symbolism, Implication, and Meaning

The concept of association covers two very different mental connections. The first connection is based on the ability of memory and the art to connect anything to anything else.

Sponge cake in tea and a person's literary life.

A statue of some guy wearing a loose-fitting robe with laurels on his head and world conquest.

The smell of a certain kind of soup and one's childhood.

The string of characters S O U P and a broad space of liquid foods.

These arbitrary connections and associations are the roots of what is usually called symbolism. Symbolism is essentially a call by reference, the variable name for a fuzzy region in the human mind. Symbolism is the complex process of memoria distilled into the basic act of seeing A and thinking B. It's as close to computer memory as humans can get.

The other mental connection is implication. Implicative connections are those that rely on nonarbitrary associations. "Where there's smoke, there's fire" is a statement of implication. It's not true, but it's still an implication.

Implicative associations are used to construct understanding and likely possibilities out of known information. Rather than going to a single place in mind the way memoria do, implications create flows from place to place through the mind's understanding of how things work.

Implications can be erroneous. Smoke can come from things that aren't fire. Once bitten, twice shy arises from the implication that a thing that bit once will bite again, or that something we think is like what bit us will bite us the next time. Many aphorisms are erroneous implications.

We build our understanding out of learning and experience, but we don't necessarily build workable ways of checking whether our understanding corresponds to reality.

The process of correcting the implications we make use of is called wisdom, and it's kind of rough going. We'll touch on it later when we discuss how to make art with programming.

For the moment, let's consider memory as an interconnected space of knowledge and understanding wherein symbolism allows one to jump from place to place and implication lets places flow together.

We tend to have a first-person perspective on this. We might mentally bounce from idea to idea or we might gather a batch of ideas together and see what they make together.

But if we take a step back and imagine our memories as an ecosystem wherein the acts of connection are alive, if we see them as exploring, elaborating, and evolving what we are thinking and what we are doing with what we are thinking, we see our minds as alive because we are alive.

If we take that step back and then another step back from that, we can reach some sense of what meaning is to us. We can think of meaning as living thoughts within living minds.

There is a tendency (fostered by dictionaries, unfortunately) to imagine that meaning is a small-scale thing, something lodging in individual words or symbols.

But if we look at our knowledge and understanding as alive, so that ideas and processes grow and evolve over the course of our lives, we'll see personal awarenesses that are not confinable within a few lines of text.

This space of flowing imaginative meaning is one we will be returning to later because it's a large part of where art is made and audiencing is done.

The reason we're stopping here at the moment is to see how disruptive this playground of imagination is to persistence. Anything can flow and change and evolve within this space. Nothing need stay fixed. Nothing in mind needs to persist.

If we seek to make art of persistence, we need to keep a firm grip on the uses of imagination. We need to be careful and deliberate with change.

That's fine as long as we understand that imagination and meaning are aspects of the human mind, not of a computer's processes.

Current AI uses expensive manipulation of large amounts of data to interpolate between human creations to make meaningless structures that human observers can imbue with meaning. It works by having a large library of human works translated into large-scale vector spaces and doing many fundamentally simple interpolations to create novelty without invention.

We know it's not imaginative or thoughtful because if one adds the results of its interpolations into its library the results decay rapidly, becoming less and less meaningful over just a few iterations.

Whereas if we look at the history of human art, we see that artists acting as teachers and inspirers of later artists lead to ongoing increase in meaningful possibilities of art and meaning generation by generation.

Humans can feel the lack of previous artworks and strike out on their own. Current AI cannot get fed up with realism and invent Impressionism.

Intelligent, imaginative software would be a wholly different thing than what we have now, and programming would become a very different art, one closer to parenting and teaching than the creation of aids to persistence.

But all of that is hypothetical. Even if there eventually is such software, we'll still be making and using unintelligent software because it provides us the benefits of persistence that come from a lack of imagination.

Reversible and Irreversible

Computer memory has added a concept to our everyday lives that doesn't exist in reality: undoing or restoring from a save. The conditions of our virtual work and play can be undone. Indeed, we rely on this in a variety of situations. There are a number of computer games that expect the player to try and fail at tasks many times before they figure out how to make it work. In many such tasks, the game environment is irrevocably changed (e.g., by character death). But that irrevocable change is revoked by restoring from save.

The computer environment is restored but the mind of the player is not. The player has learned not to try to jump over the spiky, explody, poisoned ball of screaming rage. Next time, they'll try handing over the demanded lollipop instead.

This combination of reversible computer conditions and irreversible real-world conditions gives us one of the other major benefits of virtual persistence: Ease of experimentation. We can try high-risk actions in a virtual environment with no real risk except personal frustration. We can play with possibilities, learn from them, and go back to try out other possibilities.

The use of computer modeling and virtual testing has eased the jobs of many an engineer and reduced the risks of many a person testing prototypes.

Framing Persistence

In the succeeding chapters, we'll dive more deeply into the practicalities and implications of programming as art. As we do and as you consider what programs you are making and using, remember that you can, if it helps you, return to the consideration of the software in terms of persistence and change.

See the way the program works to keep persistent what your users will be operating on with their evolving, changing minds. See how going back in time to earlier variations can let them implement different paths of change. Consider whether what you are having persist is what your users need to have persist. Is it offering them the opportunities to change what they need to or anchoring them in a strict progression that may not be how their minds are going?

Personal Exercises

1. If you didn't do the suggested perception and memory practices in this chapter, go back and give them a try.

Testable Exercises

1. Look up and in your own words summarize one aspect of persistence and/or object permanence.
2. Look up recent work on one or more aspects of human memory. Summarize in a few paragraphs, emphasizing those parts you find interesting as well as any that either support or suborn the thesis that human and computer memories are unlike each other.
3. Take a program that you use quite a lot and make notes of what persistence the program maintains. See what it lets you change and what it keeps persistent that you cannot alter. Describe how these choices help and hinder what you are trying to do with the program.

3

User as Audience

All Art as Tools the Mind Makes Use Of

Humans are a tool-using species, which makes tools very hard for us to understand. We make them and use them, but our concepts of make and use are themselves relative to the concept of tool. So here in a sense is an answer to the question, "What's a tool?" that isn't an answer, but which may be the best we get:

> Everything humans make and use is a tool.
>
> Which deliberately sounds like everything humans do is art.
>
> Which is a glib way of reiterating that art and tools are two different ways of thinking about what humans do.
>
> We can do a little better than that, especially if we go back and examine the idea of persistence and how it can lead human minds astray.

There is a fundamentally false model of reality that is founded in persistence. This model posits that things will always stay the way they are unless they are affected by something.

Newton's first law of motion uses this model while also showing that it is false.

"An object at rest tends to remain at rest unless acted upon by a force. An object in motion tends to remain in motion at the same speed and in the same direction unless acted upon by a force."

These statements are true, but misleadingly disingenuous. Isaac Newton was also the formulator of a theory of gravity that declared that every object in the universe exerts force on every other object in the universe. In short, everything is actually being affected by forces and therefore nothing is really at rest, and nothing is actually moving in a straight line (Relativity elaborates this and makes the math both more complex and more interesting).

The reality is that everything in the universe is having its motion changed all the time. But not all motions are changing quickly and not all motions are changing inconsistently. Down may only be meaningful close to a gravity well, but it's consistently meaningful. That consistency can be made use of.

The false model, its Newtonian formulation, and the kind of localization that brings forth local consistency give hints that a tool might be something that exerts force and therefore makes things move that were otherwise not moving.

DOI: 10.1201/9781003459866-4

The more accurate model would turn this into the idea that a tool changes how something is in motion. Ideally, a tool makes changes more or less consistently so that a sapient mind can learn to use it. This is the tool function of art.

The human mind is also internally in motion. Our thinking is a complex interaction of ongoing processes. We use certain kinds of tools to affect that ongoing internal motion. This is the artistic function of art, to affect the motion of a person's thinking, to give new possible directions to ideation.

Because each human mind is unique and learns from its experiences, these internal tools do not have consistent effects. Each mind at each time gets new experiences out of its interactions with art.

Internal inconsistency is a feature of art just as external consistency is a feature of tools.

We are a tool-using, thinking species. Using tools affects how we think and the thinking we do affects how we use tools. Consistent outside, inconsistent inside makes art and tools one unified awareness with wildly fascinating feedback loops.

This seeming duality looks paradoxical because of another false aspect to the model of a universe of still objects waiting to be acted upon by forces: Determinism.

Classical mechanics describes motion as following strict paths that are calculable given the appropriate information. Quantum mechanics says no, not really, motion is probabilistic. The chaos theory adds the fussy fact that being a little off in the appropriate information means you can be a lot off in the results.

What is true is that in carefully controlled circumstances, the motion of objects under forces is likely to be close to the paths predicted by classical mechanics. But in other circumstances, things will move in wilder and wackier ways.

This is how the divergence between tools made to affect the world outside the human mind and tools made to affect human minds came to be and why human thinking does not benefit from being more consistent and controlled.

It is possible to create carefully controlled circumstances in small areas of the real world (such as a worktable with paper and pencils). It is not possible to do so in the human mind. This is a feature of humanity, not a bug. The diversity of art and the usages we can make of art and the variance of people's reactions to art mean that one work of art can help prompt a panoply of reactions from the different people who experience it. This means that a carefully made drawing shown to a wide swath of people can inspire many waves of radically different ideas. That is a highly impressive range of effects for the work needed to make one drawing.

This makes our tools and art efficient in two completely opposite ways. The better the tool is at affecting reality, the more predictable its results will be. The better art is at affecting a human mind, the less predictable the results will be.

This paired difference between the external and internal effects is also fundamental to our understanding of art as tools and tools as art.

The tool user/audience member picking up and using a single tool is transforming the world around them and their own mind by the use of the tool. This ongoing double transformation is one of the basic awarenesses that an artist/toolmaker should have when thinking about their audience.

It is also an awareness any audience member should consider before interacting with a tool/piece of art. Each time a tool/artwork is experienced, the mind will generate an implicit paired question, "What can I do with this?" and "What can this do with me?"

Artworks That Are Interactive Tools

These questions are not only paired, but they also iterate in time, feeding back into each other. If you use a tool and observe its effects, the observation becomes one of the things the tool has done to you.

So, what will your mind do with that observed effect?

There are many possible answers, some of which we will explore in the next section.

For now, let's consider what happens if a tool maker works with the awareness that the user will experience and examine the use of that tool and their usage of it to make things work better for the user.

What happens if the tool makers make tools that make use of this feedback between usage and user?

What if they make interactive tools?

And what, to bring us back to the main subject of the book, if that interaction allows for adjustments in the tool itself and how it is used? What if the tool has the capacity to easily incorporate adjustments based on the usage?

In other words, what if the tool use changes the persistence of the tool, and what if one is making tools using an art that's really good at persistence?

Tools do get adjusted by usage. The wear patterns on shoes show how the shoe wearer walks. Roads get worn down by driving on them and need patching. But this adjustment isn't deliberately built in, it just happens through usage. Most tools aren't made with adjustable persistence.

Software can have a flexibility that most other tools do not because coding is the best art at creating and modifying persistence.

So, let us delve into interactive tools from the user's perspective and see what that tells us as both users and makers.

Let's focus in on the immediacy of action and feedback.

When using a physical tool, actions take time and there is a sense of feedback in the process from the feeling of doing the action. Sawing wood, for example, has feedback into the hand that is sawing. That feedback gives a practiced carpenter a sense of the texture and the hardness of the wood they are sawing through. There is also a sense of whether one's hands are guiding the saw or being pushed aside by the interior of the wood, and sometimes one can feel whether the saw is at risk of breaking.

Driving a car can give a feel for the ground one is driving over based on how the car is reacting, how rapidly wheels are turning, whether there is skidding, and so on.

In both of these cases, there is also a longer-term feedback that gives consideration to whatever next action one might be taking. Where to cut next? When and how far to turn the wheel?

With software, feedback is more likely to appear after the user has done an action. The user does something, the program feeds back, and the user examines what they've done. The user may wish to continue with the same action, tweak what they just did, undo it, or jump elsewhere. Unless the program is a game, all of these adjustments should be made easy by the programmer. In a game, the style of play determines what should be easy and difficult.

We'll delve in detail into the discussion of interfaces in Chapter 12. For now, the important thing to look at is the diversity of your possible audiences and how different kinds of action and feedback select for who can and can't use your programs easily.

Start with considering what kinds of feedback help you use a program. Also, see what feedback distracts and what positively pains you. Feedback and action depend on the kind of stimulus and response, as well as the intensity of response, complexity, detail, compositional ability, and skills of the user. There is also absent feedback, where the user does something, and all the responses are happening inside the program without anything showing up in the interface.

Every feedback structure amounts to a demand on the user. Feedback (or lack thereof) asserts an importance in what is showing up. Ideally, the feedback should tell the user what they need in order to be able to interpret the consequences of their actions. From that what they need to able to figure out what next action to take in order to do what they are trying to do with the program.

And herein lies the need to blend your artist and user awareness while also keeping them distinct. People have a tendency to make the kind of tools that they think would work for them. They'll consider what feedback they would like to receive and in what form they would like to receive it.

But if you do that too much, you will end up with a tool that only you can use. This is fine if you are making it for your own use. Personally customized tools are quite useful. But you need to be aware that that's what you're doing.

If you are making a tool for others to use, you need to denormalize your own responses by taking away the assumptions of what's intuitive to you, what's annoying to you, and what's necessary for you to get anything done. Different minds have different responses and different needs.

Different ways of showing data will work for different people, as will different forms of user action. This is a process that can grow in complexity and confusion because there are large numbers of people with large numbers of perspectives and large number of different approaches and needs. It will take time and practice to find how to broaden one's work to fit a broad enough audience space for what you are trying to do.

We'll be elaborating this in later chapters.

As you gain a clearer sense of your own methods and other people's needs, this process will become less confusing. What's needed at this stage is to remember to question your own assumptions of how things need to work and to listen to other people's reactions to your work and what does and does not work for them. It's good to have a sufficiently broad space of people testing your program.

Over time you should be able to develop a sense of which methods you employ that work for you but not for many other people. You should also eventually develop a sense of which methods work well for large spaces of people even if they wouldn't work for you as a user.

Art you make for others is not the same as art you make for yourself. The features of art for others are not necessarily the same as the features of art for you.

No matter what you make, do not let yourself be pulled too far away from making things that make sense to you. Don't let fashion drag you into doing art that is nonsensical or incoherent from your perspective. That way leads to work that will take away your ability to make art and tools that help anyone.

A good practice to go with this is to try to articulate what does and does not work for you in software you are using, what kinds of action and feedback help you and what kinds get in your way. If you can do that, you may come to see what form of feedback you can make the most use of in order to make your work more helpful to others.

Work, Play, Learn, and Share

I like to look at audiencing as a braid made of four purposes that a mind can turn to as reasons to interact with art. I also sometimes think of these as four rooms in the human mind. In each room, the mind examines the art and its uses relative to a single broad purpose.

Work: This is the workroom that looks upon other arts as tools and asks, "What good is this in what I am doing?"

This is art taken in to help the mind make its own art. This is audiencing for inspiration, or to gather materials from, or to directly incorporate into one's art, or into which one's art will be directly incorporated. Work audiencing is the most tool oriented.

Play: The playroom asks, "How can I have fun with this?"

This is art taken in for the effect it has on the enjoyment of the life of the audience. It is art as a giver of life and augmenter of the fun aspects of reality. There is a tendency to equate this with being entertained, but that is too passive a view of play. A mind playing with art will turn it around and inside out and wonder what happens if you spin it. Most fan art and fan fic arise from the enjoyer of one person's art playing with it and seeing what they can do with it.

Learn: The learning, or classroom, asks, "What of this can help me know more, understand more, and become more capable?"

This is art taken in to help teach the audience knowledge, understanding, or practice. Audiencing for learning involves using the art to create internal awareness that works for the mind that is taking it in. How and what is learned from art is mostly a matter for the audiencing mind. What it can learn from the art depends on how well the art fits the learning mind.

Share: The sharing, or storeroom, asks, "Who would get some good from this?"

This is art taken in so that the audiencing mind may make use of it in interactions with others. As a room, sharing is a storeroom or library, a place where one keeps art in order to turn it over in one's mind and find a way that others can make use of it.

These four usages/rooms are not sharply divided. It is easier to learn what one can play with and easier to work with what one is sharing, and so on. All four can be present in a single person's experience of an artwork. It is not at all unusual to experience an artwork or group of artworks that one enjoys a great deal and plays with and then is inspired to learn the ways of in order to make one's own art that is like unto what one enjoyed.

The rooms are useful as consideration when considering the creation of art because their questions can be turned around. How can my art be useful, enjoyable, learnable, and shareable?

All of which braid together into an artist's awareness of the ways of audiencing.

The Many Ways an Audience Can Follow

What is easy for one person is hard for another. What is enjoyable for one person is suffering for another. What is obvious to one person is bewildering to another. What is fascinating to one person is boring to another.

There is no one way that works for all. There is no such thing as perfect and universal. There is no universal tool. But there are toolkits.

There is no one piece of art that will appeal to all. But there are art spaces.

It is quite possible to create toolkits that will be of use to many people. It is quite possible to create art spaces in which a wide variety of people can find what they enjoy.

If you make a toolkit or artspace, it will be easier to expand it as needed than if you try to create a single tool or artwork that will work for all.

Even a single-purpose tool can be a toolkit if it allows for different ways to use it. The simplest and oldest human tools all have different ways to grip them and move them. Different cooks use the same knives differently based on what works for them. Different musicians play the same instrument in their own personal ways.

All humans develop their own ways. We learn from each other, and we adapt what we learn to how we think and act. Trying to copy someone exactly without adapting their methods to one's own is a sure path to misery.

When coding it is tempting (for reasons we will get into later) to create a linear path of action that requires the user to do exactly what you the programmer think should be done. This arises from what seems obvious to you.

But obvious to one is not obvious to another. Your work should derive from what you see as obvious, but should not presume that others see as you see. You need to make your obvious benefit your audience, not frustrate and constrict them.

If you make your work broad enough in usage, you will find users doing things with it that you never expected. This can be amazing and enjoyable, as you see new works, new ideas, and new directions emerging because you gave them a capability they did not have until the tools you made got into their hands.

Diversity outside is the fruit of consistency outside and diversity inside.

The diversity of human thinking and art is one of the unalloyed joys of toolmaking and artmaking. To see what others do with your work or to hear someone playing with your concepts and applications can be an abject delight.

To make that possible, you will need enough flexibility in what you make for others to apply it in their own way.

We each live in our own augmented reality that we create from our own minds and the factual reality we all exist in. The unique world exists on top of the shared world. We make our own ways in the shared world using our unique worlds. This creates feedback loops of evolution where we are changing the shared world for each other by making art that is flexible enough to work in more than one world.

One useful way to look at this is that each person makes their own way in the world, but we can all help each other by providing tools that make making that personal way easier.

Please note that this expansion of possible ways does not mean that your tools need to be overloaded with features. It means that whatever features there are should be graspable in multiple ways and usable in ways that let the user's ways come through. A single process that must be followed without deviation is an automatable process. Unvarying sequences of action are exactly those for which computers and robots are best suited. It is absurd to try to push those tasks on humans for the ease of the computers.

Any action that a human undertakes needs to have space and flexibility for the way human minds work and human actions flow.

The above description looks like an artist, not audience perspective, but the making and selecting of toolkits and artspaces is an audience action. We find and gather the tools we need and the artworks we benefit from. We fill up our own workrooms, playrooms, classrooms, and storerooms.

The making process and the selecting process are not separate and the finding of one's own ways is not separate from the laying out of possible ways by others.

One may never find another person whose ways are really close to one's own, but by wending one's own way around, through, and alongside those of others; one can gather and find what works for one even if the ways one uses are wholly unlike the ways anyone else does.

Rhythms and Polyrhythms between Humans and Artworks

Human hearing perceives repetitive sounds differently depending on the rate of repetition.

If the cycle is very slow, we don't notice that it's repetitive. If the same sound happens once every 17 years, it might be spotted in long-term records and it might remind someone who experiences it of something that happened to them half a generation back, but it wouldn't make a direct impression on the humans who experience it.

We hear repetitive sounds that come fast enough to create persistence as rhythms. The measurement for repetition is cycles per second (or Hertz). Repetitions below approximately 20 Hz are heard as rhythms. In rhythm, we hear each sound distinctly but are conscious of the timing between the sounds. That awareness is what makes sound rhythmic to us.

With repetition that is much faster than 20 Hz, we lose the sense of each sound and instead hear the repetition as tones or notes. Middle C, for example, is around 261 Hz. The region around 20 Hz sounds buzzy to most ears and can be quite annoying.

For most hearing people, rhythm attracts and focuses attention. This is why things like dripping faucets can be so distracting, because they bring unwanted rhythms into our minds and draw attention to themselves.

Humans seem to make use of rhythms at some basic mental level. Actions with rhythm tend to be easier for people to do. We also coordinate actions more easily to rhythms. This is why dance is so good at showing cooperation, because it uses actions coordinated to rhythm. And the world is full of rhythms. Listening quietly, it is possible to hear all sorts of repetitive sounds going on.

There are two basic musical implementations of rhythm, monorhythm and polyrhythm. Most European and American music is monorhythmic. In this kind of music everybody is singing, playing, and/or dancing to one rhythm. Their actions are taken according to a single pattern of beats.

Polyrhythmic music has multiple rhythms happening at once. Different musicians are playing to different rhythms. But you can't just combine any rhythms to make a polyrhythm. The rhythms have to exist in specific mathematical relationships with each other. It is common to have polyrhythms that are relatively prime, so that the beats do not interfere with each other. Then when common multiples happen the different players are suddenly creating great resonances together.

Each person enjoys and works well to different rhythms and within different polyrhythms. In creating arts that work in time, such as programming, it is important to pay heed to the rhythms one is imparting to or imposing on the audience.

One of the classic methods of making video games more difficult as the games progress is to speed up the action until the player can no longer adjust the rhythms of their actions to match that of the game.

Rhythms in software use show up in a lot of ways. How rapidly do screen images change? How far do mice need to move in order to do a sequence of actions? How quickly does the program require the user to orient to the contents of a new window?

Be conscious of rhythms when making software. Be aware not only of the monorhythm of the program but of the polyrhythms of the user's life. The user may be doing a lot of things at once and needing to listen for what other people near them are doing. They may have more than one program going at the same time and can't be too confined to the rhythm you are setting up.

And don't forget, what's a comfortable rhythm for one person may be too fast or slow for others.

It is generally a good idea to let your users be able to tailor the rhythms of the program to their lives and internal rhythms. Let the human set the speed and the computer attune to them.

Frustration, Confusion, Annoyance, and Rage

People have their own internal rhythms and get frustrated when events happen too fast or too slow for those rhythms. More broadly, each person has ranges of what they like, what they tolerate, and what drains and hurts them.

People also have prejudices which they sometimes treat as if they were their real tolerances. A bad reaction to something does not always mean that the person was really hurt by it. It is possible to learn to tell the difference between the two but it's tricky and people are often biased in what they accept as real or think is bias.

In any case, there are a number of possible reactions to having to deal with things outside the ranges of one's tolerance. Understanding and examining these is useful for audience and artist alike.

Frustration is a common reaction. It's actually the largest problem because it can pervade all attitudes toward computers and software. At the present time, many pieces of software walk people through step-by-step sequences of action requiring the entry of various pieces of data in order to reach a single goal. Very often, something will go wrong in this sequence and the person will need to start again. This is frustrating.

This frustration does not just arise from the specific step-by-step, but from the experience of having to follow procedures exactly without any ability for a human to modify them in order to deal with a circumstance not thought of when the procedure was created. As we said above, human reactions are diverse inside. Consistency outside should not require consistency inside.

Linear paths look easy, but they are always created by what seems obvious to one person. They never fit the fact that reality is not linear, and no person's life or circumstances fit a simple structure. Some people like following specific procedures, but even those people like to be able to tune the rhythm of the procedure to their liking.

There is a related frustrating experience that shows up often in games when there is a single obstacle to get through with a single method that may be outside of the player's capabilities, such as having to be fast or accurate enough to win a boss fight.

This kind of frustration happens when the user has to deal with a choke point that the programmer did not see as a choke point or thought was an interesting challenge that would be enjoyable to all users.

Some people like this frustration, but not many. A program that relies on this kind of frustration is a program selecting for that narrow band of users.

Confusion, by contrast, occurs when the user is confronted with the need to take action but does not have a sense of what possibilities are open to them or what the consequences of trying them might be. Confusion can arise from poorly placed or organized tools, such as menu commands in obscure or inobvious menus, or confusingly shaped buttons, or special combinations of finger taps or clicks, and so on.

Confusion is usually best dealt with by having a certain uniformity of tool use across all aspects of a program or even across most programs. Attempts to deal with confusion are how standards arise. For example, the select, cut, copy, and paste framework of tools is prevalent at the moment because the ability to do these actions crossed over into nearly all programs written to help people create things.

Confusion can happen with a clash of obviousnesses. One mind can see possibilities that another mind cannot, whether through lack of training or different perspectives or fundamental differences in thinking.

Possibilities need to be constructed and illuminated for people to be able to access them. Learning what is possible, what is workable, and what is enjoyable for a given person is vital to overcoming confusion.

Annoyance is a built-up reaction usually caused by dealing with too many things that are outside one's tolerances. Annoyance is portable and can accumulate from multiple sources. As it grows, people become more and more sensitive so that something that would usually not bother them can be more than they can handle.

Annoyance can arise when a person is having difficulty with something, and those difficulties are ignored or treated as not serious. The late Douglas Adams, author of the *Hitchhiker's Guide to the Galaxy*, did an excellent job portraying the annoyance that arises from dealing with relentlessly cheerful but unhelpful software.

When creating, one should be aware that users may not be approaching what one has made with a fully open and attentive frame of mind. It is unwise to demand that a user put everything they are capable of into dealing with what one has created. It should be possible to use most software without needing to do every action without any error at all. And, it should be possible to backtrack and correct mistakes without needing to start over.

Frustration, confusion, and annoyance are the results of moderate levels of stress, and usually manifest in a mind that is capable of taking action to deal with them one way or another.

But humans can also find themselves beyond their ability to handle situations where tolerable but painful gives way to intolerable.

Rage is one term for this state of mind, though not all minds experience rage as an explosive outburst. Some minds rage by shutting down.

Programmers need to be conscious of the phenomenon of rage quitting, wherein a user quits a program because they can't do another action without exploding or giving up.

Sometimes rage quits are accompanied by damage to hardware. More often people simply never come back to the program.

But rage quitting is only one possibility in the spectrum of responses currently called "fight, flight, freeze, fawn."

For a long time, this was called fight or flight, which was based on an overly narrow psychological model that posited that the only possible reactions to the intolerable were to attack it or to run. Freeze was added with the realization that people can also simply be paralyzed, incapable of thinking of an action to take. And fawn was added when it was pointed out that people might try to appease whatever they were dealing with.

Rage and quitting are both fight and flight responses. Freeze happens when users feel incapable of trying anything. They will sit staring at screens, not considering the possibility of trying any of the available options or asking for help.

Fawn can manifest as the opposite of freeze. The user struggles to find the command or data to enter in order to mollify the confusing bundle of hardware and software in front of them.

None of these responses are good ideas, nor do they work well. They cannot be fully prevented because people will react individually and uniquely. But it is quite possible to make one's software seem to be approachable and helpful and safe to experiment with. This won't stop the FFFF reactions, but it will make it easier for the user to come back from them and try again.

The most generally successful methods are those that make it easy for the user to see the program as their tool or toy rather than as inscrutable and judgmental.

There have been many attempts to bypass these responses by creating automated assistants, but these are only helpful in ways the programmers have already thought of, and more often than not lead the user back into frustration, confusion, annoyance, and rage. It's better to build the visibility into the program itself rather than creating a helper program that has its own inscrutable, judgmental, but cheerful problems.

Starting Out: Blank Page, Filled Screen, Tutorial, and Too Much Data

One of the most difficult times most people have when dealing with any tool or artwork is beginning to use it. The mind comes in with a combination of ignorance, assumptions, and whatever clues are presented from outside sources. At best, the mind has a cluster of secondhand ideas of what they are about to be interacting with. At worst, they have wildly misleading assumptions. In the middle is complete ignorance.

All through the history of art and learning, many different approaches have been tried to solve this problem. None of them really work. Ultimately, a unique mind and a shared artwork meet and begin to work upon each other.

As with all such situations where no method will necessarily work, there are many methods that might work. An artist can present whatever methods they wish, but each such method implicitly selects for the audience can make use of it.

Unfortunately, the audience that can get through the beginning of something is not necessarily the audience that can make use of the thing thereafter.

Here are some of the classics for software.

Blank page: Giving the user an empty canvas or document to start to work on. This can work if the program does something that the user is likely to have encountered before. A new word processor can give a blank page and the user can start typing and then nose around for how to do the various things they might wish to do.

Filled Screen: Pretty much the opposite of a blank page. The user is presented with an environment that tries to draw them in so that they begin interacting with the program. Very common for games.

Tutorial: Often a compromise between blank page and filled screen. A tutorial presents a step-by-step learning of basic tools and methods. The problem with tutorials is that they assume the user thinks a certain way and learns a certain way. They suffer from the step-by-step problem discussed above. Tutorials are often endured in frustration and often don't teach what a particular user needs.

Too Much Data: This was the original method before the concept of user friendliness evolved. In that bygone era, before anyone used a program they read the manual. These were bulky physical books that explained the capabilities and commands, sometimes with examples, often just bland dry text. Modern manuals are usually online and have hyper-links to bounce around. But they're not much better.

Too much data is too much to be taken in all at once. Too much at once doesn't help the user make the program work for them. Too much data can be organized more carefully and given clear, and hopefully interesting, introductions that point to helpful directions, followed by well-organized chapters with a lot of useful cross references and enough illus-trations. In short, too much data works best when it's a well-written book.

But this just changes the programmer's problem into either learning another art or getting someone to do the writing for them. We'll cover these and the related options in Chapter 15.

Ultimately, all forms of starting out are awkward and people need time to get to know what they are dealing with, to gather in what might help them and be able to make mis-takes as they learn without the program pushing back too hard.

The process cannot be made as good as one might like, but it can be made helpful and maybe enjoyable for some users.

Helping Users Explore and Find Their Ways

Users are motivated to use the software to do what they need or wish to do. But they are also not infinitely patient, nor should they be expected to be. They will need to put in the time and practice to learn to use their new tools, but that learning should be interesting and enjoyable, rather than a slog. They should also not feel like they are being forced to follow particular paths regardless of whether those paths work for them.

The learning itself should be worthwhile. It's not enough to learn a piece of software that will be outmoded in a few years. The work itself should ideally expand the user's ideas and awareness. Particular tools and methods should be interesting enough in usage to get the user to create and consider possibilities beyond what is presented to them.

Here are a few useful principles:

Listen to users.

Put needful things like changeable defaults in easily accessible obvious locations.

Give the user time, but whenever possible, do not make the user wait.

Always give the user more than one way to do things. Different methods spark different associations in the user's mind. Giving them multiple ways to do similar things will let them build their own ideas and connections.

Fuzzy ways are better than narrow paths. Ways to do things that can be easily tailored to user actions and styles (e.g., the pressure settings on styluses) will make each tool in the kit inspirational rather than frustrating.

Make it possible to get at some of the inner workings of the tools. Art programs that let a user create their own brushes are a good example. The tool as initially presented should be understandable and easy to use on the outside, but if the user opens it up they should be able to make more radical variations that fit their needs and act as a venue for their own artistic views.

Make the user's actions perceivable but not distracting. The user should be aware of what they are doing as they are doing it, but it's not a great idea to make this take their minds off what they are doing or what they might do next.

Build in the ability to take a step back and the ability to step forward elsewhere. This principle is common in art programs and many games. It allows for a view of the entire situation the user is dealing with (a full picture or map), and then lets them focus back in somewhere else than where they were.

Let them take notes or annotate, possibly also doodle and/or record audio. A program should be at least as useful as a piece of paper. Someone working or playing by hand on physical paper can make their own notes as they see fit, which can help them later. People have an easier time if they have references they created. Personal notes, doodles, and brief recordings that are connected to the work they are doing on your software will make things easier when they return to where they were before. Your ability to help them find what they are up to will be vastly augmented if you let them take part in the process as well.

And last and first, listen to users.

Personal Exercises

1. Take some of your favorite programs to work and play with. See how they handle feedback and what aspects of their methods you enjoy, which you rely on and which still bother you.

2. See if the four rooms model works for you. Consider an artwork relative to working with it, playing with it, learning from it, and sharing it. If the model doesn't work for you, modify it or scrap it.

3. Find a relatively quiet place, and if you can, sit quietly for a few minutes to listen to the different rhythms happening around you. If you are hypersensitive, don't do this for too long.

4. When stressed beyond your ability to handle things do you tend to fight, flight, freeze, fawn, or respond differently? If different reactions happen in different situations, consider what sets off which responses.

Testable Exercises

1. Document forms of feedback from a program you use because it's necessary for work you do. Make note of which forms you rely on and which ones you wish would go away.

2. Describe in detail your reactions to a form of feedback that really does not work for you. Make note of what senses it affects, whether the rhythms of it trouble you and so on.

3. Make note of the rhythms of action of at least one program you use a lot. What are the cycles of repetitive actions? How often do you need to click or tap in order to take actions, and how carefully must you time your actions?

4. Document at least one personal instance each of frustration, confusion, and annoyance you've had with software.

5. Document your reaction to at least one program using a blank screen, filled screen, tutorial, or too much data opening.

4

Programmer as Artist

Res and Dictum

The work of an artist redefines what practicality means. Everything humans do inside their minds and out is art. Therefore, broadly speaking, everything a person does within and outside their minds in order to make art is practical.

But because each mind is unique, what practices are beneficial to art making will vary from mind to mind. This chapter serves largely as a framework to think about what might work for you as you develop your art and evolve your awareness.

We'll begin by introducing a couple of archaic Latin terms, res and dictum. Res literally means the thing and dictum means what is said. The res of an artwork is the work as it exists in the artist's mind. The dictum is the created object. The res is only experienced by the artist and the dictum is what is experienced by the audience.

The learning and practices needed for making art consist of interweaved res practices and dictum practices. The existence of and need for res practices are largely ignored; so, most artists tend to create their own res practices without realizing that they are doing work, needful work. As we will see, res practices can look like one is doing nothing at all or doing things that are other than art making.

As the artwork is being made, the res and dictum both evolve. Because the mind is constantly bubbling in a chaotic contemplation of what it is experiencing, every step of making an artwork can transform the mind's awareness of the art, which will in turn affect what actions the artist takes next to bring the work into being.

This braided feedback process is a vital feature in the making of art, as is the fact that making art takes time. The artists living their lives are constantly exposed to new possible sources of inspiration which can add new depth or direction to the res, which in turn will affect the dictum.

The res and dictum grow and change together, and so the practices of each need to be done with regard to both. Every res practice illuminates and changes the dictum, and every dictum practice creates and reveals new possible manifestations to the res.

Res Work: Research, Design, and Organization: The Art Envisioned

Terry Pratchett talked about the "better writer in the back of my head." That back-of-mind writer has all the good ideas and has a clear sense of what the work is and how it goes. That better writer/artist does most of the building and organizing. The artist in front of

DOI: 10.1201/9781003459866-5

the mind and the artist in the back are two aspects of the same process. The front-of-mind work is sensory and easily perceivable. The back-of-mind work is nonsensory and difficult to perceive but does a great deal of the work behind the scenes.

There are a large number of aspects of human life that work like this. Digestion is one of the most illuminating. With our current understanding of molecular biology, we know that the primary function of eating and digesting is to provide the materials and energy needed to do the work of constructing new cells and chemicals that will let us continue to live.

We know that our respiratory and circulatory systems are also involved in this process, and we know how some of them are involved. We know a fair amount of why we eat what we eat.

But none of that knowledge comes from direct experience. Our direct perceivable experience of digestion is eating with our mouths, swallowing, feeling stuff happening in our guts, and excreting what is not used. We also experience breathing and feel some aspects of circulation, mostly our hearts pumping.

All of these macroscopic processes matter and have a great deal of effect on how we live our lives. We identify food by our tasting and mouth-work on it. We think about it relative to our personal taste. We have made the arts of cooking and eating relative to this.

But we do not have a direct sense of nutrition and what we need to eat or why.

Our actual digestive system is a complex underlying interconnected infrastructure that does much on a small subtle level that aggregates and harmonizes to more or less keep our bodies going.

The back of the mind is like that.

The front of the mind is like the mouth and stomach. It focuses on taste and feel, on the experience of eating, not the nutrition needed for cellular survival and operation.

Res work is about the nutrition and utility and the ways the art works on and for the humans experiencing it.

Dictum work is about taste, texture, and eating.

The two are fundamentally connected, but not in obvious ways.

In consequence, many res practices and methods do not have obvious causes.

For example, the need of a human body for iron is inobvious. Aerobic glycolysis requires oxygen to help get a lot of energy out of eating sugar, and oxygen is breathed in from the atmosphere (or, for many sea creatures, taken from oxygen dissolved in water). However, to transport oxygen to all the cells, there is a need for special red blood cells with hemoglobin molecules with which oxygen can easily bond. The bonds need to be easily breakable, and the cells are carried through a fractal branching circulatory structure to reach all the cells. The molecular bonds use the iron in the center of the hemoglobin molecules (so that we transport oxygen in rust). Therefore, we need to eat some iron, but we don't have the means of eating iron ore or iron bars and, hence, we get it from eating the bodies of other living things with iron already in them which is just as well because that's what we do anyway. These days, we also have digestible iron pills because we understand this process.

The above long paragraph is what the res work that the front of the mind trying to do looks like. It's a confusing, elaborated structure that still misses nearly all of the important information and does not give a clear sense of how to build, elaborate, or sustain the infrastructure.

Res (the singular and the plural are the same) are basically made of ongoing processes that are needed to create the artwork. Res does not look like the artwork/dictum. It really doesn't look like anything. It's an evolving infrastructure of processes with feedback between them that connect to the dictum in what look like very strange places and consist almost entirely of means to obtain inobvious materials and tools for inobvious but vital uses.

But just as with eating and digestion, we can learn to do the relevant front-of-mind practices that help the back of the mind build the res of the art if we pay heed to the basic considerations of hunger, nutrition, taste, and overall health.

And yes, we're going to push this metaphor for all it's worth. We will eventually leave it behind because all metaphors, like all other aspects of art, are only useful in limited circumstances.

Hunger, in this context, is the feeling that flows into the front of the mind that the back of the mind is in need of something in order to make what it needs to make.

It might need knowledge it lacks; so, the hunger would be hunger for research.

It might need energy to work with. Then, it might need rest. It might need to do an enjoyable activity that seems to have nothing to do with the art.

It might need to be near people who need the art to get a better sense of audience.

It might need to talk to people who have done similar or related work and get a sense of how they are going about it.

It might need inspiration for how to organize and structure the art. It's hard to tell without trying.

It might need to talk to someone it trusts about the work it is doing in order to get some outside perspective.

It might need to be audience for something seemingly unrelated that will work its way inward and help develop the needed new understanding.

Just as a person can, if presented with possibilities of food, determine what they are hungry for, the front of the mind can consider possible actions and experiences and see what seems sweet and savory to the back of the mind.

This process of considering possibilities and gauging the back-of-mind's reaction is a practice that over time the whole mind together can get better at, largely by washing away the distractions and assertions that a mind can generate to trip itself up.

These kinds of self-sabotage are regrettably common.

For example, if one is working on one kind of art and needs to experience a totally different art, for example, playing a game while taking a break from writing a book, a thought along the lines of "That can't be right" might crop up, as the front of the mind can be distracted by the idea that things must have simple, direct connections.

Nutrition is what the back of the mind needs to build and modify the res. It is what the back of the mind is getting from the experiences it is having the person pursue in the examples given above. Understanding why your mind needs certain things may not be obvious when doing the res work, but can often be discovered later.

I've often had the retrospective awareness, "Oh, that's why I was reading that book" or "listening to that music" or "needed to go shopping in this particular grocery store."

Looking at the work I did after doing these things, I can see that I was looking through likely sources where realization or experience could show up in reality so I could incorporate them into what I was making. Sometimes, what I was doing was an experiment to see what moving through a space in a certain way would feel like.

There is the complication that res are not distinct the way dicta are. The back of the mind can be working on one complex interconnected thing that radiates out into projects with little seeming connection between them. As I write this book, I am also:

Finishing writing an SFF trilogy.

Coding a complicated program.

Working on a patent application related to the program.

Making a rules system for a TTRPG.

Considering the character I am playing in another TTRPG.

Working on adventures for a third TTRPG.

Considering what I will be cooking this week.

Enjoying having had a nice breakfast.

Enjoying sharing the house with my wife and younger child.

Looking at the wallpaper on my laptop which is a picture my wife painted.

Looking around my study at various books and bits of statuary.

And a lot of other odds and ends, all of which are feeding or being built by one big res that's been evolving in my mind for years which expresses into all of the above dicta and the dictum you are reading right now.

But all this inchoate and abstract nutritional awareness is pretty deep in the back of the mind. Res work has to grow toward the front of the mind in order to be of more direct use.

That's where *taste* comes into it. As res grow, they get closer to the practical questions that eventually lead to dicta. They start to produce possible organizations and design ways things might work in fact and human experience.

At that point the front of the mind gets more involved, planning out possibilities, considering implementations, and so on. Research can become more directed. Possible designs for the dictum bubble up. Eventually, the question of how to do what needs doing appears, and from that the organization of the making flows out.

Different minds have tastes for different methods of this phase. Some people like to research by diving into multiple sources and swimming from reference to reference. Some like to focus on a few sources, each mined fully before moving on to others. Some design in depth. Some doodle or simply let ideas wash through their minds. Some people organize with outlines and schedules. Others pick things up at the times they seem to need working on, then put them down until the next bit of res work is done.

These are matters of taste. Imposing one person's methods on another person is the same error as insisting that the food one likes is the food everyone else should like.

Furthermore, taste and nutrition aren't everything. Here is where we run into the *health* problem. For most people, their backs of the mind are not really good at keeping track of the health of the body. The connection between the back-of-mind artist and the mind's awareness of the body can be tenuous. There is a region of the brain, the insular cortex, which essentially processes bodily awareness. It has connections to memory, taste, and feeling. But it's not great at diagnostics and seems to give most of its information to the front of the mind rather than feed it into the better artist.

A growing res can take energy and time away from keeping alive and healthy. Ideas can pop up at any time and demand to be cared for and nurtured without any regard for how the person is doing or what they might need or enjoy.

This is where we leave the nutrition metaphor behind. At the point where the front of the mind is conscious of the artwork as having needs and that those needs seem to be but are not arbitrary.

We could enter the more common metaphor of art making as being like pregnancy or caring for young children. But we won't go far into it because lots of other artists have made this comparison and lots of parents (including artist parents) have glared at them about it.

Dictum: Materials and Techniques

Dictum work can be generally described as composing an artwork by using the techniques of the art to combine and shape the materials of that art.

In a practical sense, which art one is doing can be determined by examining the materials and techniques being used. From this perspective, we can see that the overlap of materials and techniques leads to the overlap of arts, as well as the ability to transfer knowledge and understanding from one art to another.

But we are going a little fast because we don't have a cogent meaning of materials and techniques.

Some arts have what appear to be obvious sets of materials: Paints for painting (but what paint is is tricky), clay for pottery (but which clays do you use? and don't forget glazes which are related to but aren't quite the same as paints) and glass for glassblowing (except glass is made by the techniques of glasswork which uses things like sand as materials (and sand is used for sandpainting which isn't quite painting)).

Some arts have very broad categories of materials. Drawing can be done with anything one can draw with. In other words, the materials of drawing are defined by the techniques. Clothing-making is done with any material that can be shaped to move with a human body draped in it, and so it is defined by usage. Sewing can be done with anything that can be threadlike on anything that can be stitched together. Sewing and clothing-making overlap because many things that can be stitched together can be shaped to be moved if worn on a human body.

Writing is done with languages as materials but so is conversation. But writing is something of a misnomer. Writings do not need to be written down to be made by the art of writing. We know this because skilled writers can make dicta out of thin air (thick air can interfere with breathing). The art of writing preexists writing down and has been around for so long that we have no idea how long humans have been doing it.

Programming also has languages as materials, but those languages are computer languages. We'll talk about this in detail in Chapter 6.

Techniques are usually even harder to examine outside an art than the materials.

First, they are hard to examine because what someone unskilled in art thinks are the techniques of the art are often illusions caused by not knowing what the artist is actually doing. Neither the act of painting by putting paint on a surface nor the act of drawing by pulling an implement across the surface, leaving a trace behind, are techniques of painting or drawing. The techniques lie in what the hand and eye of the artist are doing as they guide the brush or the drawing implement.

Second, techniques are hard to examine because many techniques are preparatory rather than depository. Many, probably the majority of techniques are done to gather, extract, blend, and make the materials ready before the materials are brought into contact with the dictum itself.

The world is full of things that can be turned into materials in different arts.

There are minerals like malachite and azurite, which can have the shiny stones within extracted from the matrices in which they occur. These shinies can be carved for jewelry or smashed to flinders to make pigments which then are mixed with different media to make paints. From rocks to paints to pictures painted of the rocks (one could use malachite paint to create a picture of the rock from which the malachite came).

And all of that is simple compared to the complex preparations used to make ingredients for cooking. Getting what one needs from life is a wide array of techniques because life (as we will discuss later) is dense and complex and oh, so diverse.

Part of seeing the augmented reality of the art-filled world is seeing what can be done with what abounds in the world in order to make the art that augments it.

And third, techniques are hard to examine because every art also has techniques that are used so that the artist can figure out which of these diverse materials and available techniques to use to create a specific dictum. "What do I do now?" is the implicit question of every moment of art making. If an artwork is going along easily, the question is not articulated but is implicitly being asked by the front of the mind.

There is a recursive aspect to the discussion of materials and techniques. Tools are used by techniques to extract, combine, and apply materials. But as we've said, every tool is itself also an artwork. In order to make art we make use of the arts made by others.

Another part of the augmented reality we live in is that we use made art to make new art. As we use tools, we both acclimate to them and become critics of them. Standing in the pivot between artist and audience, the tool maker/tool user is both evolving their art and evolving the tools used to make that art.

As programmers, our tools are programming environments which are themselves programs running on computers. There are also computers that we will also use to run our software when we have made it. We can't program without computers, and computers are not computers without programs. We seek to make our programs portable/device independent, but for them to work they will need to fit into the particulars of each system they are running in.

We are dependent on the memory, processing power, and speed of the systems our software works in. Those considerations shape what programs we can make.

These kinds of factual considerations exist in all arts. At some level, all dictum work comes down to the physics of the situation.

Potters struggled for centuries to try to make very large ceramic sculptures that would not shatter in kilns. They found the limits of this for the various clays they were working with. Thermodynamics and materials science could not be ignored.

Painters learned how different media and pigments aged over time. They saw the aging of older paintings and learned the difference between fugitive and nonfugitive pigments.

Writers learned the way different kinds of writing surfaces (paper, vellum, papyrus, fired clay, and carved stone) aged over time. They learned about the differences in the survivability of scrolls and bound books, and they learned to their cost that storage environments mattered a great deal. More books have been lost in neglect of being copied than were ever deliberately destroyed.

And, of course, languages and cultures change and, hence, what an art depicted or what a tool aided in doing could be lost simply by the change of meaning over time.

Dicta are always creations of their time and context because all artists work within their time and context. Materials, techniques, and observed meaning are tricky to preserve in the ongoing passage of time and the practicalities of physics and chemistry.

This is not a reason to use fugitive materials or fill your work with pop culture references that will lose meaning within a few years. Just the opposite. If you take care of your materials and techniques, your art will more likely be preserved and learned from.

This is hard to see with an art as young as programming, but it is an art where the techniques are very visible, so we can see what we have inherited from those who came before us (including our younger selves). Examining old code shows clearly what the programmer was doing or trying to do. Your work is the code and it can teach later generations, and inspire them to learn from it and it may inspire crosswise. We do not know what arts will arise from programming in a century or so. We therefore cannot know what of our work will be of use to our successors nor what use they will make of it.

If you make your dicta mindfully with an eye to the benefit others may glean from the code itself as well as the effect of running the programs, your work may feed later generations of artists in ways you can't know but might well be pleased to find out.

All Art Has Mistakes: Bugs, Debugging, and Habits of Error

The existence of the res as a mental immanence can create the illusion that the artist has a vision of a perfect form of the artwork, but that whatever dictum the artist creates somehow fails to live up to that vision. This is made worse if one exists in a culture of perfectionism.

There is a recurrent myth that if an artist could wish the vision of the dictum into reality, then they would achieve a greater, purer work than they create by their actual work over time.

This is at best an error. At worst, it's a cruel lie that makes artists hate their own work.

The fact is that a res guides the making of the dictum but is not like the dictum. The sense of a perfect vision is not actually accompanied by perfection. The res is abstract relative to the dictum. It is not made of materials or techniques. It is a mental, not a physical object.

In a culture of perfectionism, the mental is often seen as purer and better than the physical. Again, this is an illusion. The res exists only for the artist and is of the artist's mind. The dictum is made in reality and therefore can be shared with other humans and can be used by them. The res is of no use to anyone else than the artist, but the dictum can be.

As the dictum is being created, the res can elaborate, becoming deeper, more subtle, and more complex, and extending further into awareness. As the res evolves, the dictum will be able to draw upon this evolving internal awareness.

A dictum that could be wished into being or created by the push of a button would not benefit from the ongoing work of the artist's mind and actions. The chaotic flow of new ideas in a working mind is a feature, not a bug.

However, this does not mean that a dictum is free from error. All artworks have errors in them. But error is not a deviation of a dictum from res. Error is the presence in art of something that distracts from or detracts from the experience of the art, or the absence of something that makes it harder for the audience to make use of or enjoy the art.

Error, therefore, is relative to audience, since what matters to one person will not matter to another.

Most arts can gloss over errors pretty easily. There is a saying in cake decorating that whipped cream covers a multitude of sins.

Programming has a harder time with errors than other arts because an error in a program can cause a crash, preventing the program from working at all.

For this reason, we have the term bug and the idea of debugging. Our errors are both more annoying and easier to fix than in other arts. It's much easier to fix a mistake in a program than in a painting.

Because our errors are both serious and easy to fix, finding and removing errors is a vital phase in the process of creating our art. This has the advantage that every programmer knows that they will make mistakes and they will need to debug. In other arts, it can take a lot of work to get an artist to accept that they've made a mistake or that making mistakes is something that happens to everyone.

We have a weird advantage, which is that debugging is a process that combines the frustration of searching out one's errors with the relief when one finds them and the joy when one fixes them. Few programmers are ashamed of the errors they find and catch. They may be annoyed at the fool who made the error in the first place, but they will happily let others know that they found and fixed that fool's mistakes (that fool so often being me).

Most other arts could do with learning this phase from programmers. Debugging as a concept is valuable, and relief and joy at fixing one's own mistakes could benefit many people.

When we look deeper at error, getting past the basic crash-causing mistakes, we can spot certain common patterns and kinds of errors. These need to be watched out for because they aren't necessarily seen as errors.

Three broad categories recur – errors of flattening, erasure, and substitution.

Flattening errors come from treating something complex and diverse as though it were simple and uniform. Stereotypes are flattening errors. Treating everyone in a group as if they all behaved the same and treating periods of history as the same or cultures as the same are instances of flattening. Replacing complex laws of physics when they affect outcomes with simplified versions that ignore vital parameters can mess up modeling.

Flattening errors need to be corrected at the roots. In programming, this often involves replacing a simple object or function with a more complex one. We'll look into this in depth later.

Erasure errors involve treating something extant as if it didn't exist. This kind of error often shows up as a precursor to flattening errors. Removing relevant parameters makes it easier to replace complex functions with simple ones. Depicting objects without shading can make 3D figures look 2D. Erasure errors also remove whole groups of people from the histories of places, pretending that the places were not inhabited when they were, or pretending that only some sets of people were living there when there were many different peoples.

The most common kind of erasure error is ignoring infrastructure and environment. People have messed up farming in areas by assuming that what worked in one place would work elsewhere. Not understanding the differences in climate or terrain or ecosystems leads to this kind of erasure. People have built homes and towns while not paying attention to what would be needed to supply the people there with food and water. Erasure of what is needed is very dangerous to human health and survival.

Substitution errors occur when the artist has a very superficial knowledge of materials. A cook who does not know that table sugar (sucrose) behaves differently from honey, glucose, isomaltose, allulose, and various artificial sweeteners is going to be very badly surprised when they try to bake or make candy. A painter who does not know that different kinds of red paint not only look differently and age differently but also blend differently is also in for a surprise.

The facts of materials need to be learned in order to understand the consequences of swapping one thing for another. The same applies to tools. Different methods and means of blending produce very different effects. Whisks are different from forks for whipping eggs, as are mixers and blenders.

All three of these error types are caused by overconfidence and lack of knowledge about what one is working on or with. One of the reasons for research as part of res work is to be able to avoid flattening, erasure, and substitution errors.

There are also recurrent kinds of errors that are more personal to the artist.

Idiosyncratic errors. Every person's art is unique and their methods are also unique, which means that each person will also have their own personal recurrent errors that usually annoy the heck out of them. Be aware that, yes, there are mistakes you make that no one else seems to, but everyone you know also has their own personal recurrent errors. Don't worry about it, just work to spot them, and correct them.

Habitual errors. As one develops one's ways of work, play, learning, and sharing, one is likely to develop habits that interfere with these. People create interruptions that distract or drag away from work, take the joy out of play, substitute ignorance for learning, and distort what and how one is trying to share.

Most of these can be seen as reiterations of flattening, erasure, and substitution that pop up within one's idiosyncratic ways.

Removing these habits generally involves spotting when they happen (or having someone who can see what you are doing and cares about you to help you spot them), then slowing down the process and going carefully step by step to see when and where they are happening, and carefully replacing the habitual action with a useful one.

Unfortunately, there are also habits that involve moving too slowly and hesitantly. Those need to be fixed by replacing the overly slow action with a smoother often swifter motion.

These are two basic practice paradigms, slow down to be more mindful and speed up to be more present. They may not work for you at all or may only work for some situations. While bad habits recur for everyone, the methods that work to remove them can be idiosyncratic/dependent on the particular ways the person thinks.

This leads to the most important thing to be aware of: Not all things one thinks are errors actually are errors.

Some things one has difficulty with are artifacts of how one thinks and cannot and should not be changed. This may mean that one cannot usefully use a common technique, or that an activity that is presumed to be universal does not work for one.

That's okay. If something doesn't work for you, find another way. There are many more materials, techniques, and tools available than most people are aware of. There are always other methods to fulfill various needs.

Most importantly, if your art does not work for a particular person or set of people that does not mean your art is somehow wrong or in need of correction. There is no arguing about taste, so you shouldn't have to make things for the tastes of people your art does not work for. Sometimes what is needed is to find another audience.

Layers, Drafts, and Updates

Artworks are not flat. They extend in space and time, and they transform in space and time. Artworks need to be built up in space and modified in time. The spatial buildup is layering. The temporal modifications exist on two different time scales: Drafts and updates.

Layering is easiest to see in painting, wherein a painter may start with a set of perspective lines and then put on layers of underpainting before finally putting on a top layer. Some painters (Turner is usually used as an example) almost sculpt the paint to produce real 3D effects that blend with the colors and lighting to create shading.

The colors used in underpaintings are there to make the whole work more vibrant and deep. They are often in strong contrast to the colors that will be on the top layer. An underpainting can look wildly wrong if one is imagining it as if it were the final work.

If artists have a well-developed res for painting, they will know what they are planning to do on top of the underpainting and so be able to see it relative to where they are going. Someone not familiar with underpaintings is likely to be confused by what they are seeing.

Layering occurs in many arts. In some, such as puppetry and animation, the first layers are things like concept art from which the puppets and models are derived. In programming, layering usually starts with coding the most interior processes or objects and builds outwards until it reaches the parts of the program that directly interact with the user.

Testing, refining, and debugging these interior processes can be partially done within each layer, but the layers will also likely need further debugging when later layers are put on top and when the objects and processes are incorporated into more complex or user-interactive objects and processes.

Drafts are mostly a phenomenon of writing and related arts. They consist of doing a version of the dictum, adjusting the res as one goes along, then going over the version and fixing it to conform to the newer clearer res (as well as removing found errors), iterating this process until the dictum seems good enough.

In doing the second draft of this chapter, I found that in my first draft I had written some of the descriptions of the kinds of errors with abrupt paragraphs and insufficiently cogent transitions. I then made such corrections as seemed needed and then had someone else look at it.

Drafts can benefit from editors (cough), outside readers, and discussions with other writers.

Coding also has drafts, the programmer iterating through the code to refine processes and get better interior and exterior results. Alpha and beta tests are used in place of editors and outside readers. The process (and resulting stress) is quite similar.

Drafts also occur in stage plays, movies, and video making, wherein not only the scripts but also the scenes and shots as enacted are revised and edited. Major stage plays have beta tests where they are performed in smaller towns or cities before trying their luck on main stages.

Drafts mostly exist within the timescale of working on the individual artwork before presenting it to the full audience. Updates, on the other hand, happen after the artwork is released and involve changes based on actual audience reactions and bug reports. Alpha and beta tests lie in a fuzzy space between draft and update.

Updates to books, especially textbooks, can happen when new information comes to light or matters that were only theorized in earlier editions are found to be true or false, so that later editions correct the earlier versions.

Tabletop games undergo updates wherein the processes of play in earlier versions are considered and the game designers rebuild the game, usually to a combination of pleasure and anger.

Software is the only field of art where updates are expected. Programmers are presumed to always be working on the programs they made and to be rolling out updates in a timescale of months or a few years. A program that is no longer being updated is considered orphaned. For most other arts, not updating is the norm. The assumption is that when a work is done, it's done.

Whatever that means.

Being Done: Polishing Dirt, Mixing Varnish, Close Enough, and Crashing

Being done is a tricky thing for many artists to determine or accept. The mind can keep working on a project, adjusting res and dictum, worrying about audiences, checking for bugs, examining edge cases, and so on.

This is a circumstance where it helps to have other people around who are good at understanding your work habits. Such people may not be able to tell if you are done, but they may see that your work looks done. If there is something really there that needs doing, you may well see it the moment someone else says it's done.

It is also possible that the person is right and what you are seeing that needs doing will not actually change things in a perceivable fashion. Years ago, a friend of mine coined the phrase "polishing dirt" for when an artist is fussing over details that no one else can really see and that doesn't make much difference. Having someone who can see if you're polishing dirt is very useful.

There are also people who actively encourage themselves and/or others to polish dirt. There is a kind of overfocused perfectionism that insists that small details must be exactly right or the entire work is ruined. They rarely distinguish between vital and surface elements, not seeing the difference between exactitude with fittings in a submarine or spacecraft and precision in lettering.

It is also possible to be distracted away from ever finishing. There is an anecdote about Leonardo da Vinci, that he received a commission to paint a picture and set out to experiment to mix the perfect varnish for this painting. The person who commissioned him correctly opined that da Vinci would never actually do the painting.

Between the minutiae of polishing dirt and the loss of focus of mixing varnish, there is a stage of doneness that can best be described as good enough. In that stage, the res and the dictum are close together. What is needed to be shown is shown, what does show shows what needs to be shown. Any attempt to improve is as likely to make it slightly worse as slightly better, and it would not be worth the artist's time and effort to try to improve it. Good enough is when it's time to be done.

When it's good enough and when the artist accepts that the work is done, things can get weird in the artist's mind. The res has been sitting there evolving in the back of the mind for however long one has been working on this project. The res doesn't vanish when the dictum is finished, but its place and function in the back of the mind changes.

The finished res becomes an experience to be learned from for later works. The res gathers up the dictum and starts seriously changing one's artistic awareness. Yes, this kind of change will have been going on while the dictum was being worked on. But now the res has nothing else to do but be what the mind has learned from working on it.

I call this condition crashing. I personally try to avoid doing too much during these periods. Crashing can change a lot of practices and can temporarily take a bunch of one's standard methodologies offline.

Some people get depressed in this condition, others are ecstatic. Some rest, some celebrate. I don't know if anyone simply stops working and keeps doing things in the way they have without major change. It's possible. Humans vary a lot. Be aware that this time period (which can last hours, days, sometimes weeks depending on how long one has been working) can cause radical changes in perspective and can make the world look very different as new artistic awareness is being installed.

In programming metaphor, the res–dictum pairing becomes an update to one's operating system. There can be a lengthy install time while the underlying hardware and software aren't up for doing much.

Personal Ways: Each Artist Is Unique

The skills professional artists learn usually come much later than those their audiences learn. Children will likely learn how to write and perhaps some to draw, maybe some to sing, dance, and play instruments. And, nowadays, children are learning to code. This is enough to be able to make and do things that one will enjoy, and that may entertain one's friends and loved ones.

But it is a long haul from these very basic skills to making art res that express one's own awareness and understanding in the form of dicta that one might wish to share with wider audiences.

In the process of learning, students will have a variety of teachers offering various skills and techniques, usually in a mass environment in which everyone is expected to do the same things in the same way. This is a problem because it makes it harder for the student to find what ways work for them and, especially, what ways they enjoy practicing that lead them toward bettering their skills.

The other side of this is a need to express. People moving deeply into arts usually have nascent res in their minds trying to take form. What a person wishes or needs to make art about is as important to their ability to find and forge their own ways and the skills and techniques they like doing.

It is not uncommon for a budding artist to be doing a great deal of work on their own away from a classroom environment. Everyone can and should find the arts they enjoy working in even if they do not make a profession of them.

The availability of online instructional videos has made study and experimentation easier than it ever has been. Even so, self-motivation and learning can be hard for many people.

There are online discussion boards for many arts and help can be obtained there, as well as exposure to a diversity of techniques.

As for programming, there is a vast online space of people showing what they have done and how they do it. There is a lot of help available, enough that one should be able to get a sense of what one feels good doing and what interests one. It can take some time, but nowadays no one really needs to be alone when trying to learn an art.

The important thing is to try possible ways that sound like they might work for you and see what happens. Over time, a sense of one's own art ways can show up as a reflection of the ways one has been trying. What is needful is to find what excites, what interests, what flows smoothly, and what seems to make the world appear more clearly as one goes along.

It's probably a good idea to keep some notes as one goes. Human memory has problems, as we've noted. Keeping track of where and how one learned what and which things worked well and which didn't can be helpful.

This brings up one more important thing. It can be unwise to see things as a challenge. It's almost never necessary to try to master something that does not work well for you. There is always another way to learn what needs learning.

Yes, a great many important practices are difficult and will take time, effort, and repetition to learn to do, but there is no need to make things harder than they have to be. Find the methods that work well for you and refine those rather than trying to adhere to someone else's idea of the one true way of doing things.

And do not feel confined to one art. Whatever works for you and you can give the time and thought to learn can be well worth doing. Awareness across multiple arts can help all the arts involved.

Personal Exercises

1. Examine your work and study routines. See if you can figure out what things you are doing that are necessary for you to get things done that do not seem directly related to getting things done.

2. Go to an art supply store, either physically or online, and look at the materials and tools available for various arts. Warning: This can overload people very quickly. Looking at the variety of things used in arts one does not practice can be very confusing.

3. After doing testable exercise 3, look over any of your own work that involved things you did not know well. If possible, show the work to someone who does know it well. Ask them to point out any concerns they have. From this, see what flattening, erasure, and substitution you and they have found.

4. Consider your own sense of when you are done and what you use to check if you indeed are.

5. Debug something. Does it feel good when you do that? Consider taking that sense of debugging into other aspects of your life where fixing things doesn't feel so good.

Testable Exercises

1. After doing personal exercise 1, write up a preliminary list of things you do that you think might be res practices. This list can be blank.

2. While doing personal exercise 2, make note of the multiple variants of what will look like the same thing (kinds of papers and varieties of paints and pencils). Pick one type of this and research what techniques the different materials and tools are favored for.

3. Examine at least three artworks depicting something you know well. Look for flattening, erasure, and substitution errors in all three of them and document what you've found.

4. Document the sources you've learned from and the things you have done on your own as you have evolved as a programmer. Write down any undone res you are aware of that you would like to eventually make dicta for.

5

The Five Programming Actions as Techniques

All Programming Can Be Seen as Five Actions

It's useful to consider the techniques of an art in terms of how they help the art do what it does best.

The techniques of programming fall into five categories, each necessary for the creation, mutation, sustaining, and presentation of persistence.

The five categories are storage/retrieval, calculation, branching, looping, and input/output.

Understanding these techniques as methods of persistence can help form an artistic awareness of programming as well as develop the artistic appreciation and critical skills one can apply to the programs one uses.

Awareness of techniques puts the mind midway between the needs of the res and the practicalities of the dictum. In this pivotal space, it is important to see how the way you personally do things will affect what emerges. It is also important to see how the methods that work best for you are those that will leave open the most interesting possibilities to help the back of your mind evolve what you are making for the benefit of your audience and the development of your own capabilities.

Storage/Retrieval

Storage and retrieval are the core actions of persistence. To store is to place somewhere where things will persist. To retrieve is to draw the persistent out of storage in order to make use of it.

Data storage and retrieval have crucial differences from real-world storage and retrieval. If I carry a box to a cupboard and put the box into it, the box is no longer in my hands. If the cupboard already has something in it, I'll need to remove that before I put the box in. If I come to the cupboard and take the box out, the cupboard is now empty. If I come to the cupboard with something else in my hands, I'll need to put that down somewhere before taking the box out.

None of this is true with data storage. If I put the virtual box in the virtual cupboard, whatever was there before is replaced with the box. But the box is also still in my hands. I can stick the box in an arbitrary number of cupboards and, in each case, I am putting a copy of the box there. And I can still walk away with the box.

DOI: 10.1201/9781003459866-6

If I pull the box out of the cupboard, it's still in the cupboard, but it's also in my hands and whatever was in my hands before has been destroyed.

Furthermore, if the box is properly organized (like a database), it is possible to reach into the cupboard through the interior of the box and retrieve some but not all of the contents or change the contents without removing and opening the box.

All of these differences between real and virtual storage and retrieval are parts of the techniques involved as well as being the fodder for a lot of really cheap jokes as seen in the last few paragraphs. The distinctions and the jokes can make it easier to be aware of and consider how you might wish to use storage and retrieval.

There are also, however, aspects of storage and retrieval where real-world and virtual processes share the same concerns. Distance, size, speed, and time present problems both in reality and virtuality. These four are not separate concerns, but one multidimensional factor that programmers need to be aware of.

The farther data must travel, the greater the risk of mishap, interception, or just plain having to wait around. These are the same concerns as with physical shipments.

Computer waiting around involves much shorter time periods than package delivery waiting around, but it's still there.

The shortest distance is between CPU registers and RAM. Shifting data between these is basic hardware architecture. Nowadays, it happens way too fast to notice, but early on in computer design, people spent a lot of effort optimizing this travel back and forth.

The next farther distance is between RAM and local long-term storage (hard drives and such). This again has been optimized and sped up so that only the largest files take notice-able time to be locally saved or loaded. This has led programmers to be pretty cavalier about file size. Old-school optimization methods have become minor instead of vital considerations.

The next distance is passage through a local network, wired or wireless. At this point, bandwidth starts to be a concern, and we find another space of programming concerns: Traffic control.

The regulation of storage and retrieval within a local network looks not directly at the actions but at the ability to guide and share the space and time in which those actions take place.

This aspect grows even more vital when we jump to the second longest distance, sending data over the internet so that one can download or upload to globally distant places (also to and from satellites in orbit around the Earth). Traveling globally involves signals being sent through multiple places and having them arrive where they are sent without too much interception happening.

These internet routing problems are constantly being worked on and refined behind the scenes for the immediacy of user experience. They have some of the same problems as the local network problems (bandwidth and all), but the difficulties do not scale simply, and new complications arise at each level. Some of the solutions are found in the discussion of size, so we'll defer them for a moment.

The longest distance is interplanetary and is at the moment the province of space agencies. Receiving and sending data to and from probes in orbit around other planets or near the sun or flying out of the solar system is a very specialized field and runs into a physics problem: The speed of light. Data transmission to and from Jupiter, for example, takes between around 33 and 53 minutes. Optimization cannot improve this. Instructions cannot be sent to correct immediate problems and it is not possible to have extended back and forth question and answer sessions. Nor is it possible to fly a probe by teleguidance as one would a drone on Earth.

There is one final storage distance: Offline. This is storing data in physical hardware that needs to be connected to a computer in order to store into it or retrieve from it. Warning: Old codger routines activated.

Early data and software storage was frequently done on stacks of punch cards (one byte per card) or paper tape (one byte per short segment of tape). The largest early mainframes did not have the memory to be able to store many of the files people were using them for. Programs and data would need to be loaded in from outside, run, and then shuffled out to make way for other programs and data.

Old hard drives were very bulky stacks of magnetic storage that had to be kept safe from magnets, electricity, dust, and so on.

The evolution of external storage had three considerations: Storage space, volatility/risk to data, and portability. These led to multiple inventions in successive decades: Magnetic tape storage, floppy disks, magneto-optical disks, CD-ROMs, DVDs, flash drives, and so on. Internal storage was also evolving with these same considerations.

This shifts us neatly into the question of size in storage and retrieval.

Later (in much of Section II), we'll be taking a close look at data and how much data are needed to represent various real-world objects, but for now, let's go back to the very early times when every bit was precious.

When the question of representing text in a computer arose, it was immediately obvious that one could create a code wherein each character one might type would be represented by a sequence of bits. After a little playing around, it was determined that 8 bits (1 byte) should be used. This would give 256 possibilities which were more than enough for the printing and nonprinting characters used in typing English. The one byte = one character paradigm was incorporated into the ASCII encoding. And for a long time that sufficed. But there's more than one language and more than one form of writing. Each alphabet/syllabary/character set needed its own encoding, and it was hard to shift from one such encoding to another on the fly.

When color monitors showed up and it became possible to make and store picture files, 256 colors per pixel (aka 8-bit graphics) were what was available. Looking at early graphics made it pretty clear that this would not be good enough if programming would ever be able to store photographs or create anything realistic in terms of graphics.

As available storage space grew over time, a pressure mounted to change the encoding. The one byte/8 bit standard was too confining. Larger encodings that could hold more possibilities grew up, consuming some of the benefits of advances made in storage space. Doubling the number of bytes that can be stored and then doubling the number of bytes per character were not so much of a problem. But if you also double the number of bytes per pixel (to increase the number of possible colors) and double the resolution of a picture (double width and double height), then you are multiplying the storage size of an image by 8.

At the current time, we're doing okay on bytes per character since Unicode encodes all the alphabets, syllabaries, characters, and lots of emojis. We're also doing quite well on images, although those can still eat a lot of memory. And there are programming needs that still demand more memory than current systems can handle.

My brother David (coauthor of our science and math popularizations) is a physicist specializing in relativity. He has to code models of how the universe might work. Our universe is four-dimensional. To double the resolution of one of his models requires a 16-fold increase in memory.

Three variant techniques have evolved to deal with heavy memory requirements in storage and retrieval: Pointers, packets, and partial access.

If a chunk of data has been loaded into memory at a certain location and you need to access it, then rather than copying the data again, you could reference it by the location (i.e., point to the cupboard instead of pull out the box), which will take a lot less memory – a few bytes, depending on how much memory is involved.

This idea led to the widespread use of pointers, or references, as well as a number of complications. Because if you change what the pointer refers to (by storing something new there), then every pointer that points to the location will access the new data, not the old. If you are using pointers, you need to know whether you need the object that happens to be there (in which case, you use the pointer) or the data referred to (in which case you need the value). Some languages are easier to work with than others in the matter of pointers.

Pointers as a concept are also abstracted into concepts like look-up tables, filenames, and URLs. All of these are ways of using data-light objects to refer to data-heavy contents.

Packets are a tool primarily used to deal with the motion of large amounts of data across networks. Essentially, the big data is sliced up into small bundles of data which are shipped across a network and reassembled at the receiving end to create a copy of the sent data.

The reason that this works for data but not for other things in reality is part of the subject of Chapter 7, which talks about information theory. It can be glimpsed by seeing that ½ cow + ½ cow does not equal a cow, since the half cows are no longer alive. But you can disassemble the animation file for a virtual cow and reassemble it to make a moving virtual cow.

The breakup and assembly processes need solutions to the problem of informing the receiving system about what to put together with what and where. As with all of these considerations, a variety of low-level techniques have evolved and are still being worked on.

Partial access is a catchall term for a number of different methods, all of which rely on the concept that a large chunk of data actually represents a complex object, with parts that can be accessed for both storage and retrieval without copying the entire thing out of long-term storage.

The entire field of database management uses partial access, calling out tables and records from huge files, gathering data from them, and perhaps changing their contents and saving those changed contents.

Partial access is also used in so-called lazy loading, wherein a bundle of references are worked upon and their contents are only pulled up as needed. For example, a picture gallery might have small thumbnail images that contain links to the full-sized images still in storage. Clicking or tapping on a thumbnail accesses the stored image and retrieves it for the user to see. The thumbnails take much less time and space than the full-sized images, so retrieving them creates a sense of the full picture gallery without actually showing much of it.

Pointers, packets, and partial access all rely on the idea of creating small bundles of information that mark out aspects of larger collections of data.

The technique of creating small packets of data to convey what is needed to be known in order to access large amounts of knowledge and understanding is older than speech or sapience or multicellular life. It exists on the biochemical level (e.g., with genetic information) and may be considered to exist below that on the level of light being absorbed and emitted. The tailoring of such information to particular uses is much older than human intelligence, but we have learned and developed our own uses for it (e.g., language).

The use of carefully crafted pointers, packets, and reference metadata can immensely increase the ease of conveyance and help most storage and retrieval operations.

Considerations of speed have already shown up. Moving a lot of stuff takes a long time. Increased bandwidth and processor clock rate increase the speed of operations. Speed considerations fall, therefore, into three aspects: Hardware, efficiency, and security.

The hardware processes are a whole different art requiring other books, but learning about them can be useful to give some perspective on the challenges and interesting methods being pursued to speed up transfers at all scales. If you don't know the basics of this, there are a lot of books, classes, and online references to learn them.

Efficiency is most of what we've been discussing up to this point. How to organize the processes of large-scale storage and retrieval so the huge lumbering masses of data that are now flying through our airwaves get where they need to be quickly and easily.

Hold on a minute, security says. What's the hurry? Where are you going with that massive database of personal information? Who's going to see it and for what purposes?

Speedwise, modern storage and retrieval embody a security problem foreseen by Taoist sage Chuang-Tse around 2,400 years ago. There is a section of his book which is entitled *Rifling Trunks*. Chuang-Tse talks about a lesser thief as someone who will open boxes and trunks and swipe small things from the inside. To prevent this, a person may then lock the boxes and trunks and tie them shut. But then a great thief will come in, stack the boxes and trunks, and cart them all away.

Note: I actually quoted the appropriate section of Chuang-Tse in a computer security patent. I admit to being inordinately pleased about that.

The ability to rapidly access and transfer data means that even huge databases can be hacked and stolen entirely. They will likely be encrypted, but they can be stolen and shipped out in tiny packets to be reassembled later in a hacker's system. Or they can be carried out in tiny but capacious flash drives.

Security software, in theory, slows down the transfer process. It mistrusts, requests, and examines who's asking and where are they taking things. But sometimes, it only makes things more difficult for the people who should have access and don't remember passwords or security questions, or lack the correct device for multifactor authentication.

The tension between rapid access and security considerations is an ongoing one. Some of my own programming work is involved in this. No matter what I or anyone else comes up with, I do not expect security to ever be fully solved. Chuang-Tse's warnings will, I think, remain valid for many millennia after he brushed them down on paper.

And finally, there is time. All the considerations of speed are also relevant to time, but time has one more aspect that stems from a combination of relativity and humanity.

Each person has their own frame of reference and their own considerations of time. We're all following our own rhythms and cycles while interacting with each other and sharing the same space in which we move and act.

Programming storage and retrieval in an environment in which many people act or an environment in which one person does many things, for example, in any programming environment, needs to take account of the overlapping threads of action that may try to happen simultaneously and yet need to happen in a viable order.

People accessing the same data for retrieval is not a problem, but multiple accesses for storage or one person retrieving while another is storing can lead to racing conditions wherein data are changed between one access and another, thus messing up calculations.

Certain kinds of programs, MMORPGs, for example, invite this kind of problem wherein multiple players are simultaneously taking part in the same interaction (such as a boss fight). The program needs to be able to process the effects of actions in a succinct and composed form.

This relates to a time problem that we will cover in more depth in the input/output part of this chapter: Time in user experience. The times of the computers and the times of the users need to work together even though they are operating on time scales that often differ by 9 or more orders of magnitude.

Programming storage and retrieval to deal with time can involve locking access to stored information while one person is using it, a wait-your-turn process. This is scaled up to the method commonly used during updates of preventing all access while databases and software are overhauled and tested.

Time can also be used as an attack. Overloading a system's capacity by demanding too much of it (e.g., DDOS attacks) shows that despite our current extremely fast systems, they can be overwhelmed.

Time is always a consideration, a vital fact of our lives and the universe. It will show up in even the seemingly simplest acts of putting things into and taking them out of cupboards.

Calculation

Calculation is any process that takes data and produces other data from it. This includes the creation and modification of data structures.

Calculation is why computers are called computers, because they compute, which means to calculate.

But what computation can computers really do, and how do we use computation as a set of techniques?

At the root, computers do very simple Boolean operations, taking pairs of bits and producing other bits from them. From this, operations of integer addition, subtraction, multiplication, and division are built up.

The calculations that can be done by a computer are calculations built up from previously constructed methods of calculation. Once a method is coded, it can be used again and again in the same program. Methods of calculation persist and can be composed to create other more complex actions of calculation.

The other thing that can be built from these base Boolean operations are tests wherein results are produced and then checked against conditions. These tests are also calculations, and they too can be built up into tests of greater and greater complexity.

Every data object, regardless of how it is presented to users, is really a bundle of binary numbers; therefore, every action taken upon data is a calculation built up from calculations. Therefore, every test of data objects is a test of numbers built up from tests of numbers, and thus, every test is also a calculation.

So where do we stand at this point? We know what we can build from, and we know that we can operate on anything we have constructed, provided we properly build operations from operations we already have.

From a mathematical perspective, we are presented with the space for all functions that can be constructed in this fashion from the basic Boolean operations and data storage in bits, which means we have too much possibility. We cannot explore this space abstractly. It's too big, it grows too fast, and it's too abstract to help us produce the data structures and calculations we might need for the applications we are trying to build.

Instead, we need to face the function space of calculation from the opposite direction. Rather than ask what can we do with calculation, we can ask what good can we do with calculation. We can ask how can we begin with a need for data and generate a function that can calculate data that will fulfill that need.

We can write this as three questions.

1. What do we have?
2. How do we make what we need from what we have?
3. What do we need?

Sometimes, we have a presupplied function or method. If we have a bunch of numbers and we need the sum of those values, then we can use the process of adding up the numbers that were previously programmed. We can do that because computers can more or less add.

We rarely notice when we're doing this kind of programming with simple presupplied functions. The code usually flows out quickly.

The techniques of calculation happen when we don't have a premade method. If we're staring at the abovementioned triplet of questions and don't know what to do, we confront the hidden fourth question: What can we do?

There is no single answer to this. Different people approach these problems in different ways. It can be very idiosyncratic.

I'm going to outline one metamethod that can make it easier for a wide variety of ways of thinking to find their own ways to answer the question. It won't work for everybody (nothing does), but it's pretty broad in application and translates easily into a pseudocode.

The method starts by looking at the triplet of questions as two paired sets of two questions with a gap between them.

A:

1. What do we have?
2. How do we make what we need from what we have?

B:

2. How do we make what we need from what we have?
3. What do we need?

Each of these pairings reduces the size of the function space we are examining, letting us focus in on useful possibilities. Pairing A is talking about functions that use what we have as parameters. Pairing B is talking about functions that produce what we need as results. The pairings are looking at different regions of the function space, each with their own perspective.

Pairing A produces possibilities that look like, "We can make this thing with what we have."

Pairing B produces possibilities that look like, "If we had something like this we could make what we need."

The gap between them becomes a kind of fuzzy workspace in which what A can make and what B could use can be played with so that the mind can make connections between them. And playing with possible connections is one of the things that human minds do well. It's a large part of what comprises imagination. Imagination is a feedback loop between the front of the mind and the back of the mind, between dictum and res.

Let's take an example. Suppose we have a bunch of numbers, and we need to find their mean.

A:

1. We have a list of numbers

2. What can we make from a list of numbers?

B.

2. How can we make the mean from what we have?

3. We need the mean.

Notice that question 2, the how question looks different from the perspective of what we start with (A) and the perspective of where we're trying to get to (B).

If we look at the definition of the mean, we can see that one of the processes to calculate it is to take a sum of numbers and divide them by the number of numbers.

So, in the gap above B, we can place: Mean = sum/count.

A:

1. We have a list of numbers

2. What can we make from a list of numbers?

   mean = sum/count

B.

2. How can we make the mean from what we have?

3. We need the mean.

Sum and count are both things we can make from a list of numbers. So, below A we can write:

Calculate the sum of the list and count the number of elements in the list.

A:

1. We have a list of numbers

2. What can we make from a list of numbers?

   Calculate the sum of the list and count the number of elements in the list.

   mean = sum/count

B.

2. How can we make the mean from what we have?

3. We need the mean.

And we have our calculation. The space in the gap has our pseudocode which we can then render into almost any language we're using (more on this in Chapter 6).

The abovementioned process can be nested because complex calculations can involve multiple passages from one set of data to another set until one ends up with data that fulfill needs. If the results created by the first pair are not exactly the needs of the second pair, then we end up recreating the triplet like this.

A.

1. We have an initial dataset.

2. What can we make from the initial dataset?

B.

2. If we had the penultimate dataset how can we make the final dataset?

3. We need the final dataset.

The gap can have possibilities like

We can make medial dataset I from the initial dataset.

How can we take medial dataset I and make the penultimate dataset?

We can use the penultimate dataset to make the final dataset.

This triplet can be split into its own set of two pairs with its own gap in between as we try to go between medial dataset I and the penultimate dataset. We may end up calculating multiple medial datasets down from the top, while also creating a succession of functions up from the bottom that turn what we hope to have into what we need. The goal is to eventually close the gap by producing a set of data derived from the top which we can feed into the succession of functions worked out from the bottom.

The metamethod is a bit elaborate, but it gives a wide area for different kinds of minds to play. Those minds that find it easier to consider the data they are working with will mostly elaborate from the A pairings, transforming dataset to dataset. Those minds that find it easier to create the means of transforming data will mostly elaborate from the B pairings, linking function to function. And those minds that find it easiest to consider the interaction of data and function will likely work from the middle.

This process looks like dictum work, but hidden within it is a lot of res work. First, there's the matter of learning. The process requires having enough understanding of what one is starting with, what one needs, and what methods can go between them to be able to ask the questions and let the possibilities bubble up.

It's important for a programmer to understand enough of what field they are coding for in order to have a sense of the needs and possibilities. It's also good to know beyond that field because some methods recur in different fields and one group of people may have invented a method that others could use if they know about them. A programmer who studies broadly (but not necessarily deeply) can translate a set of tools from one field to another.

Second, there is the development of an internal personal awareness of the organization of data. Different minds arrange information differently. How and what the mind associates and connects is more idiosyncratic than people think. Shared data structures that have been developed in various fields work for some people but not for others. People often need to modify or replace those structures in their own thinking to be able to understand the field of study.

When you are coding, you should develop the kind of data structures that work for you. If, because of interoperability, you need to translate data into and out of your data structures, then create calculations that go back and forth. Your code will flow more easily, and you will be able to do more with it if you primarily stick with the structures that work for you.

To take a very simple example, a point in a three-dimensional Euclidean space can be presented as a three-element array [x, y, z] where x, y, and z are numbers. It could also be presented as a structure or object with three distinct components named x, y, and z, or length, width, and height. All of these can be translated from one to another. They are ways of naming and thinking about the same underlying concept of 3D Euclidean space.

Use the structures that work best for you in your coding and pseudocoding because the forms that are easiest for you to work with are the forms that are most enjoyable to play with and imagine the possibilities of.

Third, there is your personal awareness of functions. How do you think about them? Some people understand functions as individual actions, some as sequences of actions, some as elements in function spaces, some as abstractions from multiple processes, some as graphs, some as algebraic morphisms, and some as physical tools.

Some people of less mathematical frames of mind think of them best as people doing jobs or animals acting according to the ways of the animals (or anthropomorphic stories taken from those animals) or characters from stories or videos doing their shticks.

Some people associate functions with their own bodies in motion and calculate with dance and gesture. Some feel functions in rhythms and harmony. And some use methods I haven't listed because I haven't heard about them. But that doesn't matter. What's important is using what works for you.

Use whatever works for your imagination. The better the functions fit with your thinking, the better you can play with them in your calculation.

You may need to do a bit of mental exploring (res work) to figure out how the ways you structure data and the ways you work with functions work together. Once you have these, you'll have an easier time using the method of creating calculations that I outlined or find and make some method that works for you.

And don't be afraid to share what works for you with others to gain insight into how they think about these functions. Conversing about how things work for you can be a way that can give others insight into you and give them a framework to talk about how they do things. Don't push it on someone else or say that others should employ your methods as the one true way and don't just take other people's methods without their permission. Note: If it's published in a textbook, the author has pretty much given permission for you to use it.

If you are sharing methods, then you can let others know that there are methods they may not have heard of because lots of people feel trapped by the idea that if they can't do things in a way that they were specifically taught, then they can't do things at all.

Do not ever be the person saying there is one and only one way to do something. This is never true. And do not be the person seeking the one and only way. Share what can work and find what does work for you.

The diversity of data structuring and the vastness of function spaces make room for more possible ways to do calculations than will ever be exhausted.

Branching

Branching is the programming technique that determines what action happens next. If calculation is what makes computer programming an art about computers, branching is what makes it an art about programming.

Branching is so fundamental that we rarely notice when we're doing it. If we put instruction B after instruction A, we're branching, because we could have instead put instruction C after instruction A.

Programs are made out of branching because we as programmers are determining which action the computer takes next. This is a crucial capability that separates programming from reality. In reality, things happen at the same time and those happenings are interacting. Programming as an art relies on the ability to do things one at a time.

That is part of the fundamental artifice that makes this art what it is. We can lay out what we wish to happen in each thread of a program. One step at a time. One effect at a time.

Programming works on persistence. Each action taken is a transformation of what persists, that is, a change of the current data the program is working with and the data that is currently stored (either locally or distantly). In a sense, every step taken by every computer in the world is affecting all the data currently persisting and therefore changing all the data available for retrieval and calculation.

Implicit in the structure of programming is the question, what happens next?

Branching is both the asking and the answering of that question. As you code, you are asking and answering this question repeatedly.

Most of the time the asking is implicit and the answering explicit. You code the next line and move on. Only rarely are you likely to stop and formulate the question. At that point, the art of programming must devise what the computer needs to do next.

This may all sound overly abstract and maybe a bit like sophistry, but there is an important res practice that can be done when the question pops up. What's next is a vital question in all arts because there are times when we need to take a step back from the artwork as constructed and consider the possibilities of where it might go.

This moment is a larger-scale form of the gap process discussed under calculation. We have the art as it currently is (in the case of coding, the program as written). We hopefully have a sense of what the art/program needs to do. We need to figure out how to connect where we are and where we're going.

Sometimes we realize that we could be going somewhere more interesting from where we thought we were headed, so we change the res first, branching elsewhere in the space of possible programs before returning to the dictum and coding what we need the computer to do to bring us toward our new goal.

We may end up deleting a lot of what we coded before because we have found a new goal or a faster way. The word cruft was invented to cover the code we need to remove because we found a better way. Having learned the error of our ways, we change the way we are going and sometimes glare back in time at the path we laid out in the first place.

And that isn't always a good way to think about it. Sometimes, the path of the fool needs to be trod before the paths of wisdom can be made.

Awareness of branching as a part of life is very useful since we live in a probabilistic universe where there are always possibilities bubbling up. In a fundamentally nondeterministic universe, it is good to be able to consider possibilities and try different ways depending on circumstances, knowing that each such way is a different process of rolling the dice this world is made from.

From this abstract consideration of branching, we compose the concept of conditional branching and project branching itself from the mind of the programming artist into the programmed art.

Conditional branching is possible because, as briefly mentioned in the discussion of calculation, some calculations are tests.

In many ways, this is an oddity of the time and place of the invention of computers which was also the era that brought forth Boolean algebra.

Boolean algebra is the study of all operations possible on a set with two elements. For the moment, we'll call these elements 0 and 1. There are 16 possible binary operations on this set. Each operation determines a value of 0 or 1 for each ordered pair of values of 0 and 1.

These operations can be manifested into physical tools called electronic gates which have two current flows coming in and one coming out. Low current corresponds to 0 and high current corresponds to 1. Gates can be built (or etched into chips) for each of these 16 operations, although only a few of the 16 possibilities are actually used in chip design. Using these gates, abstract Boolean operations engender real-world operations that take pairs of bits as input and produce single bits as output.

Meanwhile (okay decades before, but mean whileish), another correspondence was brewing for Boolean algebra. Truth functional logic, a formal logic that treated each statement as having one of two truth values (True or False/T or F) and had operations called And, Or, Not, and so on, was being developed.

In the truth functional logic, the truth value (T or F) of a compound statement (e.g., Grass is green And Wombats dig holes) could be determined from the truth values of the components of the statement (Grass is green And Wombats dig holes has truth value T if and only if Grass is green has truth value T and Wombats dig holes has truth value T). The truth functional logic corresponds exactly to Boolean algebra. Mathematical logic has grown beyond the confines of truth functionality, but the basic tool remains and is still employed at multiple levels.

If we use the correspondence of Boolean algebra with electronic gates and the correspondence of Boolean algebra with truth functional logic, we arrive at an ability to use bit operations as tests for truth and falsehood (T = 1, F = 0), as long as we're not too philosophically fussy about what we mean by truth and falsehood.

From this triplet of corresponding objects (one algebraic, one logical, and one physical), we create the basic form of programmed conditional branching: The ubiquitous if statement.

```
If (test) {
Code, code, code, code
}
```

The test can be any calculation so long as it generates a 0 or 1 answer. All sorts of things can be happening in that innocuous (test). The program might be pulling data from halfway around the world and checking its contents against vast complex big data analysis or just determining if a particular number is greater than 0.

The simple if-branching sequesters some actions within the results of a test. The test itself is performed. The code within the braces runs only if the test is 1 (for true). If the test result is 0 (for false), the program skips over the enclosed.

If statements in isolation are used to make sure that a set of necessary conditions (like a file being loaded) are met before doing operations that wouldn't make any sense if the conditions were not met. These uses of if correspond to the human consideration of whether it makes sense to do something, or not do that thing and carry on without it.

It is in the next formulation of branching that if statements become close to how humans often think we think, the if-else construction.

```
If (test) {
Choice A code
}
Else {
Choice B code
}
```

In this formulation, the test is run. If the result is 1 (T), then Choice A code is run. If the result is 0 (F), then Choice B code is run.

In either case, the program continues to the next statement after running either the A or B code.

This if-else formulation fits the story of rational decision-making, the concept that humans test a situation, consider what to do, and do what fits the choice they have made.

Human thinking is not that simple, and determination of action is not that clearly codable. Human minds are up to much more complex interior operations than these binary choices. But this model does fit a way humans often think in retrospect about what they have done and why they did it.

After the fact, humans can construct rationalizations for why they took courses of action and express that post hoc construction as choices between courses of action. This is a matter of making art (specifically storycrafting) that gives a way of thinking about actions taken for the purpose of trying to make simple sense of the actions. One of the retroactive story elements is the formulation "I considered (test). Had it been false I would have done choice B, but it was true so I did choice A."

This is one of the ways that computer actions can seem more rational than human actions because computer branching uses a structure that corresponds to the stories of human rationality rather than the convoluted, chaotic, potentially insightful, and fundamental experimental processes that the human mind consists of.

This fundamental distinction in how humans branch and computers branch is part of the qualitative difference between thinking and programs. We'll go into a lot of depth on this throughout Section II.

It is important to remember that when doing if-else branching, you will not be modeling human thinking and that this can be a good thing if what you need is a single sharp determination of which way the software should go.

The if-else construction is often elaborated into sequences of tests and possibilities so that the else section contains its own nested if clauses for other tests. These can get very long if that's what you need. There are variants where one can test a single value for multiple possibilities and do a different bit of code depending on which possibility (if any) turns out to be true.

The last form of branching is called calling or invoking. Given that language, I sometimes wonder why it wasn't called conjuring and given dramatic flavoring instead of making it look like mathematical functions. Of course, I do write fantasy novels, so perhaps I'm biased.

Calling at its most basic involves using a piece of code written outside of the main sequence of actions. The main sequence is given the name of the piece of code and perhaps some accompanying data and having the code run as if it were next in sequence. When the code is finished, the program returns to the sequence. So, calling looks like this.

Code before the invocation.

I call upon CargoLumperTheInsatiableToStickTogether(bunch_of_boxes).

Code after the invocation.

The practicality of invocation lies in the ability of a programmer to make a piece of code that will be used over and over again and be able to call it again and again, each time for a new use possibly with new parameters.

The standard techniques for making efficient invokable processes are covered in a wide variety of modern programming books and most programming courses. They don't need reiteration here.

The art of invocation, however, is not much discussed. Fortunately, this is an art book. So, what's the beauty of calling out and being answered by mysterious bundles of code?

The answer lies not in a single invocation but in bundles of related interacting invocations. One aspect of the role art plays in human thinking is that we create our ideas of what things are based on what they do and what can be done with them. A cluster of methods for affecting a set of things gives an idea of what those things are.

We define paints by the ability to paint with them; so, if we can use the methods of painting with a set of materials we see them as paints. Jumping to a less-physical example, in the mathematics of abstract algebra, different kinds of structures are defined by operations that can be performed on the elements of those structures. A ring, for example, has two binary operations, usually called addition and multiplication, which follow certain axioms and have specific interrelationships. Any theorem provable about rings can be invoked and used to affect or illuminate any particular ring.

Clusters of methods/functions effectively define something that is being invoked. The methods in an API are used to give abilities to affect or interact with some kinds of virtual objects. For example, there are multiple APIs to load, play, and edit video files. Without the API, the objects (the video files) aren't really anything. They're inert data. With the functions in the API, the files can be called upon to be seen and worked upon as videos. The methods make the data something that humans can make use of.

This invocation of what a thing is by what can be done with it can be a useful guide in coding. It can give a sense of what features should exist and be offered to your users. Creating invokable methods is coding those processes that look and feel like they should be possible for the kind of object you are working with. How the methods interact with each other and with users and other codes should fit how you think those methods should be made available and interconnected in the program.

It is by invocation and form of invocation that the program will cohere and compose in the user's mind. What can they get their hands into, what will come when they call, and what will take them away elsewhere when they require that it convey them thither?

On the other hand, invocations that don't seem to fit, that work oddly relative to the overall aesthetic, or that do things that don't seem to make sense with what the program does can be disruptive to the sense of what you are making.

The more coherent the invocations, the more your audience will be able to get a sense of whether your program works for them. They will also likely be able to spot gaps in your creation and suggest new features that will make what is being evoked even more cogent. Whether you think that's a bug or a feature in the interaction between programmers and users is a matter of personal taste.

Looping

Looping is the process of running the same chunk of code over and over again. Reach the end and go back to the beginning. It is a form of persistence in time, a dynamically evolving form of continuity. While storage and retrieval are the core of programming, looping is the core of most programs. Almost every piece of software has one or more interior loops that take care of most of the processes.

Loops, because of how they work, require the most optimization and the most attention to detail to ensure that the program doesn't waste time and memory. The process of loop making is one that requires a lot of attention and monitoring during debugging. Beyond that, looping is a pretty simple concept.

The most basic practical form of looping is the while loop that resembles the If statement.

```
While (test) {
Code. Code. Code. Code, lovely Code
}
```

As long as the test result is true, the enclosed code runs again. Most while loops change the conditions that are being tested each time they are run through.

There are multiple variants of the while loop. The for and for-each loops go through a sequence of values and run the loop each time using the new value.

The while loop is a workhorse of coding. It's practical and its particular implementations are often part of the style of each particular language.

The art of making a while loop is preparing for the test, making the test, making the interior code, and making sure that the code changes the test conditions as needed so that the loop ends with things in the state you need them to be in once the loop is done. Preparation is as vital to while loops as using the results. You need to set up both what will be tested and what will be changed within the loop.

The process of while loops is often an iterated version of the metaprocess of calculation. If you start from data set 1 and need data set n, then by setting up a while loop that iteratively creates data set 2 from data set 1 and so on you have transferred the work of calculation from you to the computer. And that's what the things were made for.

The second major type of loop is the recursive loop. These are more subtle and their consequences much farther reaching.

Recursive loops happen when an invocation invokes a new instance of itself. The most basic forms of recursion are implementations of inductive definitions of integer functions. In an inductive definition, the function is defined on a value of 0 or 1. For a value higher than that, the function performs an operation on the value and the result of calling itself on one or more lower values.

So, integer multiplication can be defined as follows:

```
a * 0 = 0, a * (b + 1) = a*b + a.
```

This can be coded as:

```
*(a,b) {
If b < 1 {
  Return 0}
Return a*b + a
}
```

So, to multiply 5 by 3

```
*(5,3) = 5*2 + 5 = 5 *1 + 5 +5 = 5*0 + 5 + 5 +5 = 15
```

The factorial function is also defined inductively and coded recursively as

```
0! = 1, (n+1)! = (n+1)*n!
```

Coding the factorial function is a very common early coding assignment. Slightly more complex is the Fibonacci sequence, defined as follows:

```
F(1) = 1, F(2) = 1, F(n) = F(n-1) + F(n-2)
```

Multiplication and factorial invoke one new copy per iteration. The Fibonacci sequence invokes two copies per invocation. Recursion has to be handled carefully because it can clog up a system with too many nested invocations.

Both while loops and recursive loops are meant to be built with end conditions. The test in a while loop should eventually produce an F/0 result, ending the loop. The inductive definitions used in most recursive loops have an initial step that is eventually reached. In factorial, it is 0! = 1. If that becomes the last invocation, then the instances roll back up to the top entry.

This is to avoid infinite loops, loops that do not end. When I learned to program in the 1970s, every book on every language warned against infinite loops. We were absolutely told to make sure that our loops would be finite. We were also told that in case we had messed up and our programs had an infinite loop, we would have to force the program to quit, usually by hitting the key combination control+C.

But something changed when mouse and menu systems became standard. That something was the event loop.

An event loop is an infinite loop running as the core operation of a program. That loop checks to see if any events (like mouse motion or clicking on menus) have happened since the last time it iterated. If something has happened, the event loop processes any such events and iterates again.

There are effectively two kinds of event loops, even though they are coded the same way.

In the first kind, the program does little or nothing until an event happens. Most word processors, drawing programs, and other kinds of art programs are like this. They persist without change until acted upon by the user. Their interior data are objects at rest.

In the second kind, the program does a set of tasks unless and until operated on by the user. They persist in enacting preprogrammed changes. Most programs that play audio or video and many games are like this. They are objects in motion.

In both cases, the user's actions are the external forces that create change.

As for the warning that woe betide a programmer who did not end their loops, they would have to force the program to quit – well, now there's an event for that. Quit commands in menus or key combinations dispatch these infinitely prolonged programs casually to the garbage collector. And if that goes wrong, there will be an outside force quit option. The infinite loop, bête noir of early programming, has become the tame pet of modern times, which goes to show that more things can change than people expect no matter how many times they go around the loop.

Input/Output

The four actions we've already covered are all internal to the program. None of them need to directly interact with any users. Input/output is where the human and computer meet. They are where the persistence, actions, and rhythms of the software interleave with the persistence, actions, and rhythms of the user. They are also where the user's taste tastes the software to find out if they like it.

How then are these matters to be approached?

To begin with, we need to deal with the fact that input and output are strongly hardware dependent and that what approaches one can take depends on what means the user has to affect and be affected by the system.

Given the speed of innovation, it would be unwise to ground this discussion in the specifics of past, present, or speculated future interface devices.

That said, we can divide I/O devices into two kinds:

1. Those categories of tools (such as monitors and speakers) that most systems will be presumed to have in one form or another so that one could write a general code to interact with whatever form they happen to take.

2. Specific tools for specialized purposes that will need to be coded for with awareness of the idiosyncrasies of the tools.

The specialized tools will have specific needs for data and instructions and anyone using them will be presumed to be skilled in their use. The software you make for interacting with them will need to be shaped to work for those needs and from those skills.

The general tools will be more flexible in application. And you will be more free to shape the I/O interfaces to your needs and what you think the audience's needs are.

Herein, lies the crucial difference between the I/O and the other processes: Awareness of audience and interaction.

I/O Is Made by Shaping

The output of your software will shape an environment that the minds of the users will turn into a four-dimensional persistent spacetime that they will be able to remember between sessions and learn about and become accustomed to perceiving and remembering.

But it will likely be the input and its consequences that will give your software a greater reality to your users. Output gives a sense of what is happening, but the ability to affect things creates a sense of what is possible in the human mind. A happening that one can create or alter has a different and closer sense of reality than a happening one cannot affect.

Output shapes the space. Input shapes the shaping of space.

To do this shaping and shaping of shaping, you will need to compose for the user a multidimensional space of persistence and possibilities using the output being presented, the input the user will be able to cause, and the user's personal interests and perceptions to create something memorable, enjoyable, and useful.

This may sound like an overly complex thing to do, but as we will discuss later, human minds can get a lot done in spaces with complex compositions and contrasts. You'll be able to make use of the user's interest in what they can do with your software to help them get a sense of the space you're making.

The important thing to understand is that what you are shaping are the possibilities that they will be able to make use of. You will be able to do so by the connections between input and output and between user in action and software in reaction.

Action, Reaction, and Persistent Interaction

Your software will present some of what it is doing through the output you display. There is, of course, way more going on than the user needs to see. Software is kind of like plumbing. The user needs faucets, sinks, showers, baths, toilets, etc., and needs them to produce and take away water (and other stuff) when they need it to. But they shouldn't need to have to pay attention to the entire infrastructure of a city's water system when all they need is a glass of water.

They should know it exists and have some sense that a lot more is happening behind the scenes and under the sinks, and they should know what to do in an emergency, but they shouldn't have to monitor every pipe and bend and keep track of the source of their water at all times.

They should be able to turn the water on and water should flow out from where they expect water to flow. That's the basic action and reaction paradigm. The user acts through an input, a lot of behind-the-scenes stuff happens, and the software reacts with the output effects of the reaction.

From this, we evolve the software as tool paradigm.

This has a lot of space for variation. Consider the action of turning on water. What does the user need to do to get the water flowing? Most faucets involve direct hands-on actions like turning or lifting or pressing. Each of these has different requirements on the user and each of them can be a problem for users with different hand or arm disabilities.

Hands-free faucets can be turned on using motion detectors. Where do you put the detectors? Most of them are under the faucet itself so that a person would need to stick a hand where the water would flow before the water flows. If the detector were above the faucet it might start flowing when a person crossed its field of vision, which would waste a lot of water. More annoying would be a detector that turns on a sink the moment someone walks into the bathroom.

It's also possible to have voice- and remote-activated faucets, but at that point, we are no longer talking about plumbing as a metaphor for programming, we're talking about plumbing with programming and that's a subject for Chapter 15.

The action–reaction of sink and washing is an individual action within a fairly dense environment of different activities. People do a variety of different things in bathrooms for which they need various supplies and tools in various places and the ability to move from place to place easily, often while dripping wet. That's about where this extended metaphor reaches its limits.

The point here is that your software will be serving as a distinct environment in which the user will be doing a variety of actions and will need to be able to move from action to action without slipping on the tiles, that is, without the space you are making for them interfering with their motions.

There needs to be enough space for the user to maneuver, enough ease for them to do what they need to do, and enough but not too much anticipation of what they are likely to do so that they will not find the ongoing usage annoying.

Part of the goal of your input–output design is to make it possible for the user to do what they need to do without having to stop and check where each thing is and which handle they need to pull. They need to be able to focus on what they are doing with your software rather than focusing on what your software requires them to do in order to get it to do what it does.

Composition, Movement, and Variant Paths

Your output needs to present meaningful data to your users so they will be able to figure out what is happening and what they can and should do with it. This is one of those tricky art places. The composition you create is likely to be a combination of what you would need in the situation your user is in and the stylistic assumptions that you've taken on board from other programs or things you've learned.

This is a fine starting point, but you will need to stretch out the possibilities to have your software work for people who perceive and work differently from you and from the people who you've learned from.

Remember, as we talked about earlier, that human perception relies on motion. How a person moves changes what they are perceiving and how it composes in their minds.

There are a number of erroneous ideas about perception. One of the most persistent and important ideas when dealing with computers is about reading. There is a belief that reading is done character by character and word by word that the mind composes by adjoining each word to the previous information. This is the way computers read, but not how most humans do it. Most people look at the space the characters and words are in and form an evolving fuzzy awareness of what they are seeing. A lot of mistakes are made and corrected as the mind flows over the reading space.

There are certain arrangements of data that look like they should make sense to human eyes but rarely do, because the arrangements rely on the assumption that people read linearly word by word.

The database record format and the spreadsheet format both rely on this myth. It takes skill for people to learn to go through a rectangular arrangement of information and keep track of things. Going along a single row or down a single column takes practiced focus. And humans make a lot of mistakes doing it.

This kind of data is useful for a computer, but for humans, it's better to present the data in a more human, flowing form. This is why data are pulled from databases and then stuck into more human-readable forms like form letters, which are easier to read but can still be annoying.

Some current interface elements, like menus and arrays of buttons, use this kind of rectangular word arrangement. Users often need to slow down a great deal to find which particular input element they need to activate to get the effect they are after.

Menus and buttons were created to replace the earlier method, which was to type out commands character by character. For most (but not all) people, menus and buttons are a vast improvement because they supply the names of actions and the ability to select by clicking.

But these interfaces create their own composition, speed, and finding problems. It can be easier to type a command one has memorized than it is to search through a ribbon of non-intuitive buttons to find the one thing to click on.

Contrast of Interface Elements

We've talked before about how we tell things apart using artistic contrast. In input/output design, contrasts need to be clear and easy to perceive so that the user does not have to pore over a wide spread of possibilities.

This is not as simple as we might hope because whatever methods we use to distinguish one thing from another (words, pictures, and sounds) will be symbolic and, as noted before, symbols are arbitrary. Meaningful to one person is meaningless or differently meaningful to another person.

This is not a solvable problem. Symbolism will always be arbitrary, and it will never be possible to create a universal set of signs that everyone will immediately get.

What you can do is create a set of distinct but associated symbols that your user can learn by using. It also helps if each symbol has an easily accessible description, so the user can quickly check to make sure they're using the right thing. This is why tooltips were invented.

The question of whether we will ever be able to do without symbols in programming is a tricky one. Rapidly readable, easily activatable controls are very handy. For most people, they make usage faster and simpler than, for example, natural language processing. Speech can be slow and unwieldy. Human languages evolved for complex artistic communications, not quick actions.

Voice control is useful for many people, but voice commands need to be about as simple as visual symbols. Visual symbols are more useful for a different set of users, and haptic motion sensory controls are important for yet other people. It's good to make software that is controllable in multiple ways because people live and act in many ways and what is unnecessary or unhelpful for one person may be enjoyable and/or vital to another.

Presence

The input/output interface of your software is the clothing the interior code wears and moves in. The interface needs to be tailored to work with how the program moves and transforms data. It has to reveal aspects of the interior persistence in such a way that the user can perceive and feel the underlying processes.

This is a matter of what you're showing and how it relates to the human life and actions of the user.

Clothing reveals bodies in motion, generating an illuminating sense from the neuromuscular and skeletal actions that give an awareness of humanity in action within the environment. Clothing doesn't reveal circulatory, digestive, immune, or other interior systemic processes. It doesn't reveal internal organs. It shows the visible life that moves through and is moved by the environment, not the infrastructure that makes the motions possible.

The manner in which your program reveals its motions as part of the human world, the manner in which it helps augment human reality, is its presence. That presence is the beauty side of interface design and creation.

Presence is one of the areas where current software design lacks art.

There are three basic errors commonly made. No prizes for anticipating what they are.

1. Flattening: Listing the interior processes step by step as if for debugging.
2. Erasure: Not showing anything going on inside, simply shifting from before action to after action in a single output action.
3. Substitution: Using an animation while internal processes are going on or after they've happened.

It can be tempting to make an animation that uses the actual interior processes, showing transformations between important steps. This can work well if the interior process is a loop. For example, showing fractal growth generation by generation illuminates the actual underlying process and is fun to look at. But this is purely passive.

It's better if you show processes that the user can pause and tweak, to change the ongoing interior actions to fit their actions and needs.

If you do that, then the presence of the program becomes the presence of this user using this program for this purpose at this time. It brings the user and the software together on the level of action and reaction while an ongoing flow is happening in the software and an evolving awareness is happening in the user's mind.

Presence is difficult to cogently describe because it's a back-of-the-mind awareness that guides the front-of-the-mind actions. Tailoring presence can involve creating inobvious revelations that the back of a user's mind can put together to get a better sense of what's going on. A good principle for presence is to reveal to others what is blatantly obvious to you but few other people seem to directly get.

Five as One and One as Five

The five techniques overlap and interact and nest within each other. There are methods of interplay that are matters of efficiency, such as what tests to run, when and how to most easily calculate something, and when and where to input and output.

All of them are there for you to use as works for you. You'll likely develop methods of putting them together that fit your own ways of thinking and acting. And even if you are coding with others on a shared project, you should be able to use your own ways in your part of the job.

As you get more adept and more familiar with how you use these, you will likely find it easier to do the res work of taking a need for a piece of software that does a particular thing and seeing how to make a dictum for it with your particular style of technique use.

Because the five nest together, this will work on any coding scale from whole programs to multiple interacting methods to particular functions and data structures and down at last to individual chunks of coding.

The more you work and play with and make the techniques yours, the more you will be able to do with them for whatever purposes you are coding and in whatever languages you work in.

Personal Exercises

1. Pay attention to the storage and retrieval processes you are doing on your computer for around five minutes. You may get overwhelmed much faster than that.

2. Play a game for a little while making note of what calculations you are aware are happening behind the scenes.

3. Think back on a recent circumstance where you made what you think of as a choice between a few possible options. See if you can focus in on the before and after of the choice. In a lot of cases which way we end up going is another back-of-the-mind thing rather than a front-of-mind calculation and test.

4. Examine a few programs with an event loop that is interrupted by user actions. Make note of the timing necessary for the user to take meaningful actions during the interruptions. Consider what would happen if this were sped up or slowed down.

Testable Exercises

1. Write a pseudocode for a simple action of storage and retrieval for an every-day problem. Then consider what speed, distance, size, and time considerations might show up and see how you might expand the pseudocode to deal with those problems. Stop before you find yourself trying to solve all the problems with the internet.

2. Take a real-world calculation problem and frame it using the metamethod outlined in the Calculation section.

3. Look at a program that has a menu interface. Each entry in the menus is an invocation. See if the invocations add up in your mind to a cogent sense of what the program works on. If it doesn't, consider what you might wish added to or removed from the menus.

4. Write out a pseudocode for a recursive process found in nature (as if you were able to program the universe).

5. Take one process you wish that a program would show you and let you affect. Describe the conditions in the program where it would show up for you to change and write a pseudocode for that output and input process.

6

Languages as Materials

How Are Computer Languages Art Materials

Materials are what artists mold and shape their works from. Artists gather them, take bits of them, and arrange them as needed.

Some materials are like paints, with a shared medium and various pigments mixed in to make specific colors which can then be placed on a surface. Some materials are like clays, which can be mixed together, then molded, possibly glazed, and fired to produce a wide variety of useful and decorative forms.

Some materials are like metals to be extracted from ores, then hammered or melted, and shaped carefully with tools. Some are like stones to be carved and then brought together. Some are like wood and leather or silk and cotton, extracted from life and then extracted even more and carved and sawn and shaved and spun into other uses.

And some are like languages.

Languages as materials have extensible vocabularies and evolving grammars that allow for infinite numbers of combinations that can grow arbitrarily large. But the materials must be combined in ways that fit some form of the evolving grammar. Arts that use languages as materials are given shape around a sometimes-hidden source of meaning that can make sense of whether or not the combinations make sense.

All human languages are languages. Their hidden sources of meaning are the minds and cultures of the people and people who speak and write them. These are also the sources that coin new words and make new usages so that the languages evolve.

Mathematics is a language. Its hidden sources of meaning are logic, proof, and what the math might be modeling, and of course, the humans who are using it, evolving it, testing it, and expanding it.

Computer languages are languages. Their hidden sources of meaning are assemblers, compilers, interpreters, and the hardware upon which they run. Their meaning lies in the fact that what is rendered in them can be transformed into the data and instructions that cause data processors to process data.

Each language, like any other kind of material, has ways that shape it easily and ways that take more work to shape it. Each programmer, like any other artist, has ways that they enjoy shaping their materials and ways that annoy them. To find the languages that work for you, you should see which ones are easily shaped the way you enjoy shaping things.

DOI: 10.1201/9781003459866-7

Low-Level Languages

We're going to briefly reiterate the development of computer languages from the bottom up with a focus on their material qualities.

The first languages are the machine languages, which are distinct for every design of a processor. Machine languages send direct commands to the CPUs of a system in sequences of bits (i.e., actual electric signals). The signals have the chips store, retrieve, and do basic calculations. Using calculations for tests, the machine languages can branch.

At this level, the distinction between data and commands is nonexistent. A number that is also a command can be stored, retrieved, and executed.

The distinction between data, command, and address is also nonexistent, so that an address can be stored and retrieved and what is stored at the address can also be stored and retrieved. A stored address can itself be used to store a command, retrieving first the address, then the command from the address, and then executing the command that allows the machine language to loop by going back to the command stored at an address, and going on from there.

Machine code has no safety and security built into it and hence no protections against error. Everything that can go wrong is available to go wrong.

For most people, working in machine language is a nightmare. The binary sequences have no memorial value. Machine code doesn't look or feel like a language to most of us.

But there are some people to whom machine code is clear, bright, and flexible. They can shape it and they can use it and make new versions of machine language and design chips for it. There is a small set of programmers to whom this space of work is natural and enjoyable.

If you like machine code and consider it workable and playable, you might wish to pursue this vital corner of the art. None of the rest of us could make our art without you and we should all be grateful that you are doing this and enjoy doing this.

Whether you are or are not one of the dancers in binary, we need to leave machine code in the depths and slip up one level of programming materials. We're going to move machine code one step closer to human awareness and look at assembly languages.

Most assembly languages are really machine code with aliases for the instructions. The commands are assigned short sequences of letters as code words. Numbers can be entered in hexadecimal rather than binary. Addresses can be given labels in order to refer to them. The fact that to the computer command, data, and address are not distinguishable is concealed by giving human-visible distinctions that let people code in assembly language.

In order to do this, a special kind of program must be written in machine code. That program is called an assembler, and its function is to take the text of the assembly code and replace it with the machine code equivalent.

Assembly language is an important intermediate step between machine code and higher-level languages. It has almost all the same dangers as machine code, but it's harder to confuse command, data, and address. It's a bit more human-friendly, and for some people, that's all that's needed. Work always needs doing down at the base where computers are made capable of being of use to humans.

If machine code doesn't feel right to you but assembly language does, and if the problems of getting things to work at this level appeal to you, this may well be a space you would enjoy working in. We also need skilled hexers.

Once the idea of making a more human-friendly language that could be translated downward to a less human-friendly one was thought of, it wasn't long before the theory of making much friendlier languages where actual words were used and numbers could be written in decimal and much more complex commands could be executed was put into practice.

But would it be worth doing that? Machine code and assembly code both connect directly to the underlying architecture of the chips that are the reality of the computers. Why make something that doesn't have that deep connection with the reality of hardware?

The question begs its answer. If a language could be created that was not tied to particular computer architectures, then it would be possible to transport code from one machine to another, even to create software that could (with appropriate tweaking) run on vastly different machines. One programmer could make a program for many and various machines.

For humanity and portability, high-level languages were created.

Classical Languages

Before we delve into the early high-level languages, one aspect needs brief elaboration. A high-level language needs a program on each system in which it is implemented that takes the high-level code and creates a machine-appropriate assembly code which the local assembler can then turn into machine code.

There are two different approaches to going from high-level to assembly: Interpreting, which does a line-by-line translation and therefore can quickly execute small bits of code, and compiling, which translates the entire program and can do things like code optimization to make the code work better on its way down.

Interpreters are more efficient for short programs, but when you run an interpreted program you are actually running the interpreter.

Compilers take longer to run but generally produce cleaner, more effective assembly codes. Once compiled, the lower-level program can be run over and over on its own without needing to go back to the compiler each time. A compiled program is stand-alone rather than being data for the interpreter.

Depending on the use for which a language is intended, decisions are made as to whether to write interpreters or compilers for it. If one is interested in writing such programs, then it's important to learn what the differences are in how they are made and what needs each is trying to meet.

Compilers and interpreters are vital infrastructure for programming. The programmers who make them are as important as bridgebuilders. There is a hidden art to making such connections. If that appeals to you, do look into it. We'll leave this subject to people who actively enjoy the doing of it.

Now, let's break open the old libraries and look at the classics.

Early languages specialized. They experimented in elaborating their capabilities for the needs of particular groups of people or to see what could be done with computers in non-computational usages. Here are a few of the languages that were bubbling and growing at this time period:

Fortran for science.

COBOL for business.

APL for math computation.

SNOBOL for text manipulation.

BASIC for students.

LISP for list processing.

Algol for elaboration of algorithms.

Forth for alternate ways of calculation.

C for keeping low-level capability in a high-level language.

There were three directions/questions that underlay this experimentation and elaboration:

What data can the language represent and how does it represent it?

What capabilities does the language have and how does it let programmers extend the space of capabilities?

How can humans work with the language? How does it fit itself into human senses, hands, and voices?

Each of these areas of expansion contributed to the transition between classical and modern languages.

Many of these languages are still in use. Their expansion and evolution have not stopped and many are still valuable materials for those who make use of them. I sometimes tease my brother that he still codes in Fortran, but it still is one of the best languages for science.

Let's look at these three questions and see why they exist and how they pushed the evolution of languages as materials.

What Data Can the Language Represent and How Does It Represent It?

Examining this requires a review of very basic computer science. I'll be quick about it, but I'm doing so because we'll use a little of it now for basic stuff and a lot of it in Chapter 7 for very deep stuff.

So, individual bits exist in computer memory and processing as actual electrical/ magnetic or other physical phenomena. Grouping bits together increases the possibilities exponentially (for once this adverb is being used for its literal meaning).

One bit can represent one of two states, 0/False/Off or 1/True/On.

Two bits together can represent any of four states, 00, 01, 10, or 11.

n bits can represent 2^n states.

After a certain amount of playing around with possibilities, early computer scientists decided that 8 bits with $2^8 = 256$ possible states made a useful unit. They called it the byte.

One byte was good enough for ASCII to represent characters. It could also be handy if one had a short list of things.

But 256 was pretty limited if one had a large amount of data (large being relative).

Two bytes could represent $256^2 = 65,536$, which was quite nice for long lists of things.

Four bytes could do 4,294,967,296 which was pretty good for most integer arithmetic, especially if you split that into $+/- 2,147,483,648$. It could also do a decent job for colors.

But all of this is integer stuff. What about things that are represented by real numbers?

That takes a lot more data. In fact, it can take an infinite amount of data for one useful quantity. Certain vital constants, π and e, for example, require infinite amount of data in their entire decimal expression to be accurately presented.

And here is where science came to the rescue of mathematics.

At the time, programming was math. Early on, there was no separate field of computer science. When I was in college in the early 1980s, programming classes were taught by the math department. Mathematicians for software and electronics engineers for hardware were the two sides of computing without there yet being a specialized field between.

Science saved the day by bringing in a tool they had been using for a long time: Significant digits.

Science, relying as it does on measurement, came up with ways to represent the quality and accuracy of the measurements they were using. π and e didn't need to be much more accurate than the other quantities they were working with. If you were calculating the area of a circle as πr^2, you didn't need π to be much more accurate than the radius you had measured. A little more accuracy helped, but you didn't need to go very far down the decimal expression to get a value more accurate than any measurement could be.

The question of how many bytes to give a real number quantity could be determined by how accurate the measurements of that quantity were.

Low-precision floating point numbers were usually given 4 bytes. Double precision used 8 bytes and long double used 16. Over time, double precision became the workhouse of real number calculations.

So, if we look at the three things that exist in machine code, we can see that commands were put in humanlike language and data were allocated byte space based on need. That leaves addressing.

If each byte in memory had an address, then accessing an address would also take a certain amount of data. If there were only 256 memory locations, then addresses could take one byte (8 bits). If around 65,536 or less locations existed, two bytes (16 bits) would work. If around 4 billion locations, then 4 bytes (32 bits) would work, and if you needed much more than that, then 8 bytes (64 bits) would do it.

Because of binary arithmetic, computer science started using terms for amounts that are slightly off from other uses. Kilo as a prefix means 1000, but for computers, it's 1024. 64K bytes is really 65,536 bytes. Mega went from meaning a million to 1024^2, giga is 1024^3, tera is 1024^4, etc.

Early on a kilobyte was a big deal. Having a personal computer with a 64K RAM was astounding to me in the early 1980s. But as memory grew at a massive clip, it wasn't long before addressing needs grew and the standard went from 2 to 4 to 8 bytes per address.

How visible these changes were to programmers depended very much on the languages they were using. High-level languages varied a great deal in how much the programmer had to be aware of data allocation and addressing. This depended on how strongly or weakly typed the language was.

If one introduced a piece of data into a program, the computer would need to allocate space to store it. Depending on what kind of data it was, the computer would allot different amounts of storage space. Here are a few common data types:

Boolean 1 bit
Character 1 byte
Short Integer 2 bytes
Integer 4 bytes

Float 4 bytes

Double 8 bytes

Long Double 16 bytes

Pointer: However many bytes are used in an address.

A strongly typed language would require the programmer to declare what kind of value a variable would hold and require a statement declaring the existence of a variable of that type. A weakly typed language would take care of data allocation dynamically depending on what kind of values were assigned to each variable.

What that meant was that a weakly typed language would have a lot more work going on in the created assembly code than a strongly typed one, and a high-level programmer would not be able to control the underlying memory usage.

Strongly and weakly typed languages have very different feels to them. This is one of the crucial distinctions in the materiality of the language.

Most of the classical languages listed above are strongly typed. LISP and SNOBOL are not. The basic objects they work on are lists of things and strings of characters, respectively. Both of those require the ability to allocate lots of memory to variables and constants without worrying about the programmer.

One of the languages, C, is very strongly typed with a great deal of pointer arithmetic and a need for programmers to pay attention to exactly how much space each variable has and where it points to. The evolution of C, as we will examine later, is one of the oddest things that happened in the history of programming.

On to our second question.

What Capabilities Does the Language Have and How Does It Let Programmers Extend the Space of Capabilities?

All of the languages agreed on the need to expand the capabilities of what the programmer would do. As a result, all built up libraries of routines that could do a wide variety of calculations.

After all, what good is a computer that can't compute what you need computed?

Just because we can represent real numbers that are close enough to what we need for calculation doesn't mean that we can do the calculations we need with those numbers.

It took some work to code real number addition, subtraction, multiplication, division, and exponentiation to integer powers, but that work was mostly a matter of taking care of what bytes represented what parts of what numbers.

The calculation capabilities of computers might have stalled out there had it not been for a crucial bit of calculus and two mathematicians named Taylor and Maclaurin. They created a way to represent any sufficiently well-behaved function using only addition, subtraction, multiplication, and exponentiation to integer powers.

Their work is embodied in the Taylor or Taylor–Maclaurin series which is taught in most calculus classes.

Because of their work, it is possible to code trigonometric, exponential, and logarithmic functions, as well as a number of others.

Also, thanks to pointer arithmetic, it became possible to create vectors and matrices and create routines for all of these. The core operations of quite a bit of the mathematics needed in the sciences were opened up to the voracious data processing of computers.

Beyond this base agreement of coding these math functions, the languages diverged pretty rapidly. Each language created its own specialized functions and its own ways of letting programmers create new functions, each according to the aesthetics of that language.

The programming techniques we covered in Chapter 5 used pseudocodes because we didn't want to tie down to any language. Pseudocode is a kind of like preliminary sketching, which can be used for drawing, painting, puppet-making, scene setting for videos, and so on. It's not really possible to make a preliminary sketch without some awareness of what art one will be using for the final work, and it's not really possible to do pseudocode that is fully language-independent. I've been using a C-type pseudocode because as we'll see there are a lot of C-type languages out there these days.

What you are going to do in the future will influence what you do as a preliminary setup.

So, how did the classical languages go about organizing the code itself? How did they get the programmer to think about what they could get the computer to do?

Each language would essentially put one of the five techniques as the central one and treat the other four as assistants. Of these, the most common were calculation-centered and invocation-centered.

In calculation-centered languages, the primary process is to calculate values and assign them to variables which will be used in other calculations. Input is used to get values for variables and output to present the values of variables. Looping and branching are employed in the service of calculation. Storage and retrieval are implicit in the variables, and sometimes explicit in other data storage objects like the Stack in Forth.

The apotheosis of calculation-centered languages was APL, which used special characters to represent various sophisticated mathematical operations. Special keyboards were often used to enter these characters. Statements in APL are very hard to read but do a large number of calculations with very few keystrokes.

LISP and SNOBOL were also calculation-centered, but they were dealing with larger and more complex datatypes than number-focused languages like APL and Fortran. They were working with lists of objects and with strings, respectively, and were therefore delving closer to matters of human language.

The functionality of SNOBOL focused on pattern matching within strings of characters and was eventually folded into the syntax of regular expressions. Regular expressions have become a language within a language for many other languages. This is a good example of the material qualities of the languages. SNOBOL pattern matching and regular expressions have effectively the same capabilities, but SNOBOL uses words in a command form whereas regular expressions use single-character expressions clustered together. I enjoyed the feel of SNOBOL and loved working with it. I dislike the texture of regular expressions and avoid using them.

Invocation-centered languages focus on the construction of code processes (usually called functions or sometimes procedures) outside the main flow of the program and invoking those outside processes inside the main flow.

Calculation- and invocation-centered languages diverged on the matter of how to do what could be done. Calculation-centered languages lean toward finding and using clean, efficient single-purpose actions, often with minimal side effects. Invocation-centered languages open the space up toward much more complex side-effect-producing operations.

One of the most focused invocation-centered languages is ALGOL (algorithmic language). Algol is also the name of the star used as the eye of Medusa in the constellation of Perseus. The name comes from Arabic and means the ghoul. This has nothing really to do with the language, but it does give amusing/unnerving associations.

Anyway, an algorithm is a laid-out procedure, like a recipe. Invocative programming focuses on the coding of algorithms that one wishes to use in one or more programs.

The method I laid out in Chapter 5 when discussing calculation is an algorithm used to create algorithms. As that algorithm shows, the creation of algorithms can be done from a multitude of directions and encompass more purposes and usages than all of humanity will ever need.

ALGOL gave rise to the language Pascal, which expanded its capabilities and for a while was a very popular language for introductory programming classes. Pascal brought algorithmic thinking and invocation-centered programming into the foreground. It had considerable influence on a generation of programmers.

As we will see, calculation-centered and invocation-centered languages blended together and gave rise to a different dichotomy in the present time.

But first, our third question.

How Can Humans Work with the Language? How Does It Fit Itself into Human Senses, Hands, and Voices?

People vary in what materials they like to work with. Does it feel too sticky? Too runny? Can you move it around as you need? Does it hold its shape? Is it too inflexible? How does it smell? How does it look? How easily does it blend with other materials? What can you apply it to? What tools are needed to work it? How narrow are its applications?

In programming, most of these manifest in three aspects. How much work does it take to do what you need to do? How easy is it to learn and remember how to do what you need to do? How well do the methods you need to use blend into the methods you like to use?

Languages vary in how they allow the expression of the five techniques. Some have terse syntax (e.g., C is quite terse in its loop structures) and some are quite verbose (COBOL uses a lot of words for most of its commands). Verbose is usually easier to remember but takes longer to write. Programmers vary as to which of these matter to them.

Different languages have different syntactic requirements. For example, C requires that single lines of code end with a semicolon (;). Python, which is related to C, instead indents lines of code and requires no terminating character. Going between C and Python can be a bit of a nuisance because the muscle memories conflict when typing.

The amount of work involved in a particular language can get very heavy in matters of input and output. Fortran, for example, allows for very exact formatting for how lines of input and output will be presented.

Modern interfaces connect multiple input sources into the code and then the code has to connect to multiple output sources. How these are built and arranged can please and frustrate.

For example, some languages make use of prebuilt interface objects and make it difficult to create them within the code itself. Many people appreciate this process. Others find it confining. More about this in Chapter 12.

This leads to a broader area. What does the language take care of for you and what do you need to take care of yourself?

People vary a great deal in what they like having others do for them. Each person/method of doing things makes assumptions of what needs doing and what to leave undone.

If these assumptions don't meet the needs of the person being done for, then they will have to undo or find a workaround.

As we discussed before, the more the system insists that things be done in one true way, the more frustrating it will be for someone who does not work according to that way.

The narrower and fussier the critical paths to follow are, the more the language is narrowing who will like it and who will not. Defiles like that tend to sharply divide users between fans and antifans. Languages with more ways to go tend to have wider groups of people who like them because of the greater ease in adjusting the language to the programmer rather than insisting that the programmer must adjust to the language.

On the other side of this issue, languages that have clear methods for doing the basic techniques of coding are easier to adapt to because they are easier to dive into and play with and learn from.

Some classical languages had very complex requirements at the start of any program wherein particulars of the hardware the program would be running on had to be specified exactly.

So, what happened to languages as computers evolved and went from taking vast amounts of space for very little memory and computing power through the era of the personal computer to the current (2023 CE) situation of many people owning multiple portable and wearable devices, all interacting through multiple networks?

Three different kinds of languages evolved: Object-oriented, functional, and everything we do on the web starting with HTML.

Object-Oriented Languages

Once there was an experimental language called Smalltalk. It played with the concept of object-oriented programming (OOP). That is, Smalltalk programs focused on the creation of complex data structures, which gave each kind of object its own set of invocations, called its methods.

Object-oriented languages (OOLs) changed the perspective of programming from focusing on actions to focusing on what the actions are being performed on. This was a completely new perspective that takes in aspects of both calculation-centered and invocation-centered languages and says, let's look at what we're working on and what those things need to work upon them.

Object-oriented languages make use of ambiguity in intriguing ways. They let the same name be given to different functions applied to different kinds of objects. For example, in a number of OOLs, the operation + applied to integers does integer addition, to floating point numbers does floating point addition, and to strings it concatenates. Programmers can create their own objects and define a + for those objects, so in a program the line a = b + c can mean almost as many things as it might in abstract algebra (where it could mean an infinite number of different operations).

Object-oriented languages have inheritance, wherein a new object type is created as a subtype of an earlier type, inheriting all the components and methods of that type, but possibly adding new components and methods and perhaps overriding the methods of the earlier type.

Calculation and invocation were made into aspects of the objects themselves. In some ways, OOLs make objects that are like icons or statues. What the statue is wearing and carrying tells you what it is a statue of. Carry a sword, it's a soldier. Carry a scythe, it's a farmer or death. The props and costumery of the methods both determine and reveal the character of the underlying object.

While every OOL allows for the creation of new types in the creation of objects, not all of them are strongly typed. Some insist that variables have declared types, but even then, that type will also encompass all subclasses of that type. So, if Wombat is a subclass of Mammal, any variable declared to be a Mammal can have a Wombat as its value but any variable declared to be a Wombat cannot have another kind of Mammal as its value.

There are also seemingly strongly typed languages that also allow weak typing with a bit of a workaround. The language Swift, for example, operates as strongly typed but has the special type Any which can hold anything.

This useful ambiguity is also bothersome. It makes some programmers twitchy to both have to and not have to pay attention to type. Such programmers find themselves pushed away from this kind of programming. And that's fine. People should work with materials that work for them, and nothing works for everyone.

To complicate matters, each OOL has its own syntax for creating classes, objects, and methods and operates on them in its own way. The language C has a number of OOL descendants (which is not something I saw coming because of how low-level C itself is and how high-level OOLs are). C++, C*, Objective-C, Swift, and Java are all OOLs descended from C. Each has its own syntax and requirements for creating and working on objects. It's easy for one person to really dislike some of these and really enjoy others, or to enjoy different aspects of several of them and wish they could be combined in a way more convenient to the particular programmer.

The emergence of OOP is one of the things that shows programming as an art. There is no functional difference between OOP and non-OOP. Both can produce the same effects. Both can construct the same kinds of data structures and functions. But OOP has a different feel to it and a different sense of how one is shaping what one is shaping. Which to use is fundamentally a matter of the taste of the programmer rather than the needs that the programs are being written to fulfill.

Functional Languages

The second new language trend that emerged is that of functional languages. These also incorporate aspects of both calculation- and invocation-centered programming.

A functional language program involves a lot of function definitions, then applying those functions to supplied data, and the results of applying other functions to other applied data with the possibility of lots and lots of side effects along the way. The program will look like functions nested within functions nested within functions. It's a bundle of algorithmic awareness with data treated as what happens to come along.

Functional languages, like OOLs, are strongly a matter of programmer taste. To a mind that easily sees reality as processes nested in processes operating on what happens to need working upon, they can seem utterly natural. To minds that focus more on what is present or on how what is present interacts with what else is present, they can be very confusing.

OOP and functional programming feel very different as materials. They flow differently and they frame problems differently. A programmer who feels lost using one may feel as if they've come home when they learn the other.

It's a good idea for a programmer to learn at least a little of a variety of languages so they can find what works well with their ways of coding. They can also find features in a language that they don't like but wish were available in languages they do like. They might be able to grant those wishes for themselves by coding those features into their favored languages.

It's important when learning computer languages to gauge how much time and practice you need to get a sense of the language, become competent, and master it. It's also important to find out how the process of learning feels to you. If it bogs you down to learn a new language, then don't do a lot of sampling. Learn what you need to learn. If you enjoy the learning process, then cast a wide net and learn what's out there and what is emerging.

HTML: Language and Interface

Something happened in 1993. Tim Berners-Lee invented an interpreted language, HTML, meant to be easily used on a wide variety of computers. HTML was output centered. It lacked almost all of the other techniques. It had a special form of invocative branching (clicking on links) and the ability to retrieve contents from one URL to put into another page, but otherwise was almost completely incapable of getting a computer to compute.

It became incredibly popular in a very short period of time.

HTML wasn't looked at as a programming language. Its name, Hypertext Markup Language, says that it's about presentation, about creating a visual presence for interconnected text sources. Its usage is so unlike most other programming languages that a web designer is treated as a different job in the field. But it is a programming language, just one focused on what the user is seeing, not what's going on underneath.

HTML made the process of constructing output elements easy compared to all other languages. It gave interfaces a blank page to work from and tools to fill it with. But in its initial forms, it was lacking the infrastructure of storage/retrieval, calculation, branching, looping, and input.

And those lacks cried out for other languages.

Mixing Languages/Mixed Media

HTML's surface abilities can only do so much on their own. Pure HTML pages need to be set up individually and changed by hand. They produce output without much ability to handle input.

HTML can do a great deal more when connected with other arts. Like glazing on pottery or painting on architecture or clothing on dancers, the appearance is better if it's wrapped around other arts. HTML benefits from mixing media with other languages.

There are three common methods for such mixing: Code embedded in a page which is largely the province of JavaScript, calls to server-side programs that affect the contents of a page usually by accessing databases, and code that generates HTML.

JavaScript is a C-descended language that is used to give client-side capability to web pages. It can be used to change the characteristics of what is being presented, give capability to input controls like buttons, and store variables that will affect the display on multiple related pages. JavaScript essentially expands HTML to be input/output centered, not just output centered.

Calls to servers and codes that generate HTML were for a time mostly the province of PHP, another C descendant, but the space of server-side languages has expanded a great deal recently.

HTML uses various protocols (starting with HTTP) to send and receive information from servers and incorporate the received information into the page, sometimes directly and sometimes by transferring information to JavaScript (mixing the media even more).

Page construction programs use the fact that HTML files are plain text files with special markers inside them. Any language that can build a long string and save it (i.e., just about any language) could in theory create HTML. In practice, the major web development languages have special purpose procedures and syntax that make it easier to make these strings on the fly. To do so well requires that the programmer either have the skills of a web designer or be working with one. We'll talk more about this in Chapter 15 and Chapter 19.

What Is the Future of Programming Languages/Materials?

I don't know. The evolution of programming languages is a part of art history that is too young to be fully studied. And even for arts that have long and diverse histories, it is unwise to predict what might come. Arts evolve through the invention and interaction of people, and people can always surprise you with new materials and techniques and new uses for old ones.

That's part of the fun of being human.

Personal Exercises

1. Think about the different programming languages you've dealt with. There are likely to be different sensory associations in your memory. They might feel or sound or look or even smell or taste differently to you. You may experience working with them as if molding them in your hands or walking or swimming or flying through them. Cultivate this awareness. When you are coding or reading code, pay attention to the associations. If you can refine this internally, you'll be able to develop a good sense of whether a language works for you or not.

2. If you don't know them, look up the classical languages listed earlier. Read their syntax and a bit about their instruction sets. Look at examples of code. Use the awareness from exercise 1 to see how they come across to you. Make note of which ones seem compatible with your ways of thinking and working and which don't.

3. Look at any website you use frequently. Examine it as output-centered programming. See how it is putting its presence forward and its other coding elements in the background.

Testable Exercises

1. Examine the syntaxes of any three C-descended OOLs. Describe how they create classes, write methods, and create objects. Talk about how each of these feels to you and what you like and dislike about them.

2. Write one piece of a pseudocode for an OOP and then write pseudocode that does the same thing for a functional language. The pseudocode should input one piece of data, perform three distinct functions or methods on the data, and then output the result. Note how the flows of the program are similar and different for these kinds of languages.

3. Look at the HTML for any complex webpage. Document its invocations of other languages.

Section II

Too Much Information

7

Information Theory

This section of the book focuses on the practicalities of programming as art. As is standard for any practical study, we will start by digging into theory. Specifically, we're going to delve into information theory, which developed with the conceptualization of computing.

We'll be looking at two information theories, classical and quantum, because we're going to be looking at two different kinds of information and how they fit into the relationship between art and reality.

As is very common in mathematical theories, we are going to take two words that are synonyms in everyday English, data and information, and use them as names for contrasting ideas. So, yes, we are burdening your vocabularies with a distinction you will neither need to nor probably wish to make in real life but which can be immensely valuable for the art you practice.

Information theory, both classical and quantum, primarily concerns itself with the ability to efficiently and accurately transmit data. A lot of bandwidth has been spent and conserved because of this.

But what we're going to look at is what the data/information in these two theories consist of and what the contrasts between them mean for programming as art.

Classical Information Theory and Meaning in Information

Classical information theory is based on the idea that all information can be decomposed into and composed from bits. It's immediately obvious that if one wants to represent the elements of an infinite set using bits, you will need an infinite sequence of bits since it takes an infinite number of binary place entries (bits) to represent the positive integers.

Information in bit form is therefore only usable for representing finite sets of possibilities. But each of those possibilities can themselves be infinite.

If, for example, I'm comparing the integer, rational, real, and complex numbers, then I'm talking about four possible objects (requiring only two bits to distinguish them), but each such object has an infinite number of elements (integer and rational numbers having a smaller infinity than real and complex numbers). The information needed to pick among these sets is finite, but the information needed to pick an element from any of the sets can be infinite.

Given any finite set, we can easily calculate how many bits we need to represent all the elements. It's $\log_2$ of the number of elements in the set rounded up. So, 100 elements take $\log_2(100)$ rounded up = 6.643 rounded up = 7 bits.

What those seven bits provide is a lookup table for the 100 elements in the set rather than a meaningful structure of bits corresponding to the elements of the set.

DOI: 10.1201/9781003459866-9

Meaningful in this context means connects well to human thought rather than fits into computer memory. And herein lies a utility and a problem with bit-based storage.

As we know, any bit-based information can be stored anywhere it can fit in a computer.

But it's hard for humans to memorize and handle arbitrary lookup tables. Consider how long it can take children to memorize the order of an alphabet. Songs are made to help learn this (Song is a very handy tool for human memory and has been used as such since well before writing was invented). But even once remembered, most people have difficulty in using the alphabet for looking up. Someone may know the alphabet, but if asked, many would have to say "A,B,C,D,E,F,G,H,I,J,K" to answer "What's the eleventh letter of the English alphabet?"

Human minds are not great at lookup tables. But we do still use classical information structures in our everyday lives. We have made a number of tools, including tables and numbered lists, to make it easier for us to look up arbitrary listings.

Human minds dabble in classical information theory. We ask and answer meaningful questions that have only a small number of possible answers.

"Is that a goat or a hamster?"

"Have you had lunch today?"

"Do you agree with the foundational ideas of Cartesian dualism?"

It's true that every one of these questions has not only a simple answer, but a set of fuzzy answers (especially the first one, if the animal is hard to identify and might be a wombat or a therapsid or something else with a lot of fur). We'll deal with the fuzzy answers in the next section.

People can ask what they think are simple questions and expect simple answers, even if their expectations are unwarranted.

The thing is that human/meaningful questions are those that work with human thinking and human needs. They can be given the form of classical information theory questions (answerable with bits) if the possibilities in the lookup table can be organized meaningfully.

There are a number of games (20 questions being one of the classics) wherein the asker must make purely binary inquiries requiring yes/no or true/false answers in order to find out what object or person someone else is thinking of. These questions are often category based. For example, "Is the object an animal?"

This requires that the asker and the answerer share a concept of what is or is not within the category. When there is disagreement, people often enter into arguments over what the defining characteristics of the category are, which mostly reveals how little agreement there is and how fuzzy human thinking is. There are complicated disagreements over what constitutes a religion or even a sandwich.

To prevent these things from turning into violent conflicts, subcategories and other categories are created and another lookup table is created. Classical information theory continues to be connectable with meaning if the lookup tables are provided with new labels. Sometimes new academic disciplines – and cuisines – are created to accommodate the evolution of classifications.

This makes it look like the problems of human classification are solvable by classical information theory since all we need to do is come up with a set of binary questions that can meaningfully classify any particular object regardless of whether the person being asked is willing to call it a sandwich or a religion.

An argument was made by the late historian of religions Ioan Couliano in his book *Tree of Gnosis* that a number of historically recurrent human ideas could be organized into tree structures with meaningful binary questions forming branches between the different forms the idea could take. The book uses the religious concept of Gnosticism as an example.

There were and are a lot of variations of Gnosticism. Couliano pointed out that those variations could be placed in a tree structure by asking a number of Y/N questions. Here is a part of his tree reformulated as pseudocode.

```
1. Is there such a thing as a soul? Y/N
If Y {
2. Does the soul preexist the body? Y/N
3. Is the soul created? Y/N
4. Does the soul reincarnate? Y/N
}
```

Most forms of Gnosticism have Y for questions 1, 2, and 4 and could go either way for 3. But Couliano also posited that nihilism was structurally similar to Gnosticism in various ways and most nihilists answer N for question 1.

What's important for us isn't the variations of Gnostic theology (and its relationship to other theologies). What matters here is that Couliano's structure's branching provides a framework to look at what kinds of answers each branch of Gnosticism is giving to questions of the nature of the universe and humanity's place in it. In Gnosticism, the answer to the latter is usually some variant of the universe is bad and humans should try to get out of it.

This abstract tree structure view of large-scale recurrent phenomena is useful in classification and in finding underlying similarities between things. It can point to why there is a commonality of recurrence. If every form of Gnosticism effectively has to answer a bunch of yes/no questions to know what it is talking about, then different forms of Gnosticism can share underlying ideas derived independently. Couliano's trees also allow for a classical information theory way of finding where within a recurrent framework a particular idea might lie.

Using a Couliano tree, it is easy to make a set of questions that can determine where two forms of an idea agree and where they disagree. This opens up fascinating areas for historical analysis in which recurrence and its consequences become tools for analysis and comprehension.

But while that analysis can find large-scale commonalities, it can't predict or classify the artworks and manifestations that one mind or culture will produce from their local form of the recurrent idea.

The *Hymn of the Pearl* is a piece of Middle Eastern Gnostic poetry from the third century CE, existing in an evolving context of early Christianity interacting with other religions. The hymn is found in one of the Gnostic gospels and is similar in form and poetry to other writings of that time and place. It can best be analyzed artistically relative to other contemporary mystical explorations with poetic expression.

The Manicheans, a syncretic Gnostic religion, migrated from the Middle East in the early centuries CE through Persia and arrived in China around 650 CE. This migration produced art all along the way, including a number of images that draw upon Chinese Buddhist styles and iconography.

There are still Manichean groups in China to this day. Their art can be examined as arising from their religion, but also relative to the art styles of their times and places. The iconography of Chinese Buddhism (itself an interaction between Chinese art styles and imported Indian Buddhist imagery) was employed by Manichean artists both because it fit the art of the ambient culture and because of syncretism between Manicheanism and certain sects of Buddhism.

Couliano's binary analysis shows that the *Hymn of the Pearl* and the Chinese Manichean art share an underlying theological awareness.

But the form of mysticism and the expression of the religion into art that different groups of gnostics practiced are also products of the ambient culture, the methods of art being locally practiced, and the interactions of various religions at the times and places in which the art was made. And, of course, the art is always the product of the individual artists who made it, who themselves had their own interactions with the art and cultures in which they grew, learned, and made.

Suppose we are presented with the problem of creating a program that will organize Gnostic ideas and manifestations. The Tree of Gnosis lets us organize the different forms of Gnosticism relative to their ideas and attitudes. Classical information structures casually handle the binary branching and the tree takes up very little space and is easy to navigate using questions with y/n answers.

But how do we deal with the art and the way the different groups lived their lives?

The artworks themselves are very data heavy and difficult to search through.

Common solutions involve increasing the amount of classical information attached to the artworks by adding metadata tags to the files of the artworks to make them searchable using classical information searches.

Such searches amount to adding more binary questions of the form, "Does this file have this tag?"

Using this system has searches that extend through the metadata. The results are then interpreted by algorithms that determine whether or not to present each file as a possible hit, such as Gnostic y/n, Chinese y/n, and Syrian y/n.

The metadata might be further burdened with lists of similar artworks, times and places, and so on.

Metadata can get quite bulky in an attempt to give files enough ability to pin down where they fit for a given search. And no matter what, the search will eventually present the results to a human to determine which if any fit the personal human search they are trying to do.

The difficulty classical information theory keeps encountering is that the more one looks at reality, the more information one finds. And no matter how hard classical information theory tries to reduce that to bits, the more bits it finds it needs.

Why is this? Why when we dig down do we keep needing more?

Why can't we make the real world do one thing at a time? Why can't there be one bit after another? Why can't we take one discrete step at a time that does one single action at a time?

Because classical information theory doesn't fit the universe we live in.

Quantum Information Theory and Reality

Information theory developed in the early to mid-20th century. At the same time, two revolutions were going on in physics: Relativity and quantum mechanics. We've already touched on relativity and art, and we will do more later.

Quantum mechanics brought in three things that transform information theory and present us with a very different universe than the one made of bits. Those three things are uncertainty, probability, and entanglement.

The uncertainty principle states that certain pairs of quantities, called conjugates, have a funny kind of relationship as regards measurement. The more exactly you measure one of the quantities, the fuzzier the value of the other quantity becomes. The usual form of the uncertainty principle is $\Delta p \Delta q > h/4\pi$, where p and q are the conjugate quantities, Δp is the uncertainty of p, Δq is the uncertainty of q, and h is Plank's constant (which is a very small number but it's not 0).

The most well-known conjugate quantities are the pair of position and momentum and the pair of energy and time. These are important quantities in physics. If we go back to Newton's Second Law, F = ma, you'll find all four of position, momentum, energy, and time buried in there in one way or another.

Quantum mechanics transformed this most basic equation of motion from a sharp-edged thing where you could plug in exact measurements and get exact predictions to a fuzzy something that could give you okay approximations of what more or less be happening.

To get those fuzzy values, you had to be careful to remember that every time you gained exactness in one aspect you would lose it in others.

From an informational perspective, the uncertainty principle increased the amount of data that went with each measurement. You couldn't just say 5 kilometers/second for velocity, you needed to be aware of how accurate you were because of how inaccurate you would be. In classical physics, accuracy of measurement leads to accuracy of result. In quantum physics, that's not as true as one might like.

Conjugacy generated another problem for information theory. Bit-based information has a fundamental separability. Chunks of classical information can be separately generated and stored in such a way that one piece of information does not interfere with others.

This, to reiterate, is a founding principle of information architecture in computers. It's implicit in the storage and retrieval process. Store one chunk and retrieve one chunk.

But conjugate quantities are not separable and the information about them is therefore not separable.

We can fudge this by using significant digits. After all, we knew that we weren't using exact measurements in the first place. We can still use computer data and computer calculations as long as we understand that we're not accurately representing the way of the universe. We are modeling using a model that doesn't fit if we look too closely because the act of measuring closely affects more than what we are measuring.

So, what in practical terms does it mean that a quantity is uncertain?

It means that the universe is made of constantly rolling dice. Uncertainty manifests as probability. Instead of knowing where things are and how fast they are moving, we can know the likelihood of where they are and the likelihood of how fast they are moving.

This adds a complex geometric element to the information about an object in motion. We cannot say that this object is at point (x, y, z). We need to present pictures of where that object might be.

Here are pictures of electron orbitals which represent the probability distributions of where electrons in an atom are (Figure 7.1).

The darker areas are where an electron in that orbital is more likely to be. Lighter areas are less likely.

Early models of atoms required only one piece of data about an electron, its energy level, which gave a simple orbit shape. Later models required energy and two pieces of angular momentum information.

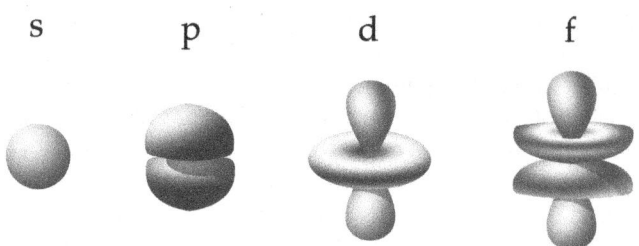

FIGURE 7.1
s, p, d, and f orbitals.

But what that information gives us is not a position or a path. It gives the probability distribution/orbital shape of the electron's uncertain position.

The energy level of the electrons in an atom determines what energies of light the atom will absorb and then emit. Those energies determine the spectrum of the atom which, upon interacting with our visual receptors, determines what color we see it as, depending on the ambient light.

The quantum numbers of the electron tell us not where it is but what complicated dice are being rolled to determine where it would be if we measured it.

That's the base information structure we're looking at when we examine and model reality. Fuzzy dice.

All of this tells us why classical information theory is only an okay fit for reality as long as we don't look too closely. Where classical information theory really stops working, where we really need quantum information theory, is the third tine in our fork, entanglement.

To understand entanglement, we need to back up a bit and talk about quantum numbers.

In quantum mechanics, each particle has a convenient list of numbers that represent what kind of particle it is and what state it's in. Each number comes from a discrete set of values. This sounds like a convenient reduction in needed data. Many numbers which in classical physics could have any of an infinite set of values can in reality have a small number of possible values. Every electron has the same rest mass (around 9.1×10^{-31} kilograms) and the same charge (−1).

Electrons also have a quantum number called spin, which represents a kind of inherent angular momentum. Electrons have two possible spins: Up (+1/2) and down (−1/2).

Why electron spins are +/− ½ rather than whole numbers is a huge, complex rabbit hole with the names Fermi, Dirac, Bose, and Einstein and a whole lot of consequences of the difference between integer and half-integer spin which we are not going to go down, but is really interesting if you're into physics.

Take a look back at the orbital diagrams. In particular, look at the s and p orbitals. The s orbital consists of a single sphere. The p orbital has three dumbbell (paired teardrop) shapes. The sphere and the dumbbells are called the lobes of the orbital. Each lobe can hold up to two electrons. According to the Pauli exclusion principle, if the lobe is full (has two electrons), those two electrons must have opposite spins (one up (+1/2) and one down (−1/2)).

Not a major problem. Two electrons are attracted and lodge into the lobe, one up and one down. If you remove one, it will have the spin it went in with, right?

Uh, nope. Once the electrons are in the same lobe they become entangled. Neither one has a definite spin. Each has a probability of being spin up or spin down and its state is said to be the superposition of those two possibilities.

It's as if each of the electrons has become an unflipped coin waiting to be either heads or tails. The coin is flipped if the electron experiences certain kinds of particle interactions. These interactions have unfortunately been called "observation."

This is one of those cases where the technical term messes up people's understanding because the word "observation" has implications for humans that imply that observers must be sentient living beings.

This has led to a long irritating confusion wherein people interpret quantum mechanics as requiring consciousness to somehow determine or choose outcomes.

That's not what's going on. Observation here is a kind of interaction on the particle level, not someone squinting their eyes small enough to see and recognize the spin of an electron.

"Observation" as a term is bad nomenclature, not deep philosophy.

Back to the coin flip. Interact properly with one electron and it will show up as having spin up or spin down. What about the other one?

We might think that since we have two electrons, we have two observations and two coin flips, one for each electron.

But that's not how it works.

There's only one coin for the entangled electrons. Flip it for one electron and it determines the spin of both. If one is observed to spin up, then the other will spin down. This is true even if the electrons are separated by vast distances.

Discovering this produced the Einstein–Podolsky–Rosen paradox (or EPR paradox), which has been a headache for physicists, a gold mine for SF writers, and an utterable joy to quantum information theorists.

Information theory is concerned with the transmission of information. The EPR paradox allows one to transmit a bit of information from one electron to another by just interacting with one.

Let's slow this down.

There are two electrons. The spin of each electron is one of two states. Therefore, each spin value takes one bit. Therefore, two electrons take two bits.

If the electrons are entangled, then measuring one of them will give the states of both of them. One bit of measurement gives two bits of information. This applies at both ends of the entanglement. If one person measures one electron and finds it spin up, then they'll know that the other electron is spin down; therefore, they will know not only what they're getting but what someone on the other side measuring the other electron is getting. One bit is transmitted and two bits are gained. That's a massive improvement over the classical one bit transmitted and one bit received.

Except it's not that simple.

At first glance, entanglement can look like it's a reduction in the information in a system. The paired electrons went from each having a state to neither having a state, so we've erased two bits.

But we've actually created a much more complicated information object. Because the indeterminate state isn't information-less. It has a probability for each of the two possible resulting states.

So, let's say the state where the first electron is spin up is $(+1/2, -1/2)$ and the other state is $(-1/2, +1/2)$. There is a probability p that when measured we will get the first paired state and a probability 1-p that we'll get the second. What number p equals is also information. If p is 0.5 or 0.7 or 0.44499999123, then the entangled state contains the amount of information necessary to represent p. The fact that we are flipping a coin with the probability of heads p contains more information than the two bits we anted up for our entanglement.

Entanglement increases the amount of information. Quantum information theory notes this and works with it. However, the major concern of the theory is the transmission of information. It uses entangled information to see about transmitting more information. The theory also uses entanglement as a way of representing the ambient information noise caused by how much entanglement is actually going on in a real-world environment.

Our concern isn't transmission; it's how does this excess information affect what we are perceiving and how does the existence of entangled information shape reality and what we can make of it. What information theory calls noise is the world around us and the world we are made of.

Quantum information theory tends to examine entangled electrons when they are no longer attached to atoms. Free but entangled electrons can fly apart and be measured at great distances. Information teleports from one to another at the speed of light. Transmission protocols love that.

Let's go back and look at those orbital shapes and their probability distributions and consider the consequences of entangling one electron to another within orbitals. What happens to shape, motion, and color?

In short, how does quantum information affect the things we use to make and understand art?

Structuring Information and the Evolution of Complexity

There are two ways electrons can get into orbits and their orbitals. The first is by the atom attracting free-floating electrons (a positively charged nucleus attracts negatively charged electrons). This only works when the electrons are moving slowly enough to be pulled in. Nuclei in stars do not fill up their orbits because the electrons are too energetic.

This fill-the-shells process keeps going until there are the same number of electrons in orbit as there are protons in the nucleus, at which point the atom is more or less electrically neutral. Electron shells generally fill up from the lowest energy on up. Within each shell, orbitals fill up in order. The lowest shell has only an s orbital. The next batch has s and p. d is added later and very late in the stacking of things f shows up.

The outermost shell of the atom will in most cases have empty spaces that could be occupied by electrons. But other electrons aren't being pulled in because the atom is electrically neutral.

Most of the periodic table of the elements is essentially a listing of how many filled spaces there are in the s and p orbitals of the top shell. Each column represents a number of electrons from 1 on the left to 8 on the right. The extra rows have to do with shells with other orbitals (Figure 7.2).

But the empty spaces in the orbitals can still be filled in even if no free electrons are dragged in. The method of filling we're going to look at is covalent bonding.

In that process, two atoms connect by entangling electrons in their top shells to form molecular bonds.

When entangled, the electrons not only lose the information of their spins but also which atom they came from. If a bond is broken, one electron will be in one atom and it will have a certain spin. Which atom and which spin are two coin flips.

That's not the major information effect of this form of orbital filling. There are two more important changes that happen when atoms bond covalently.

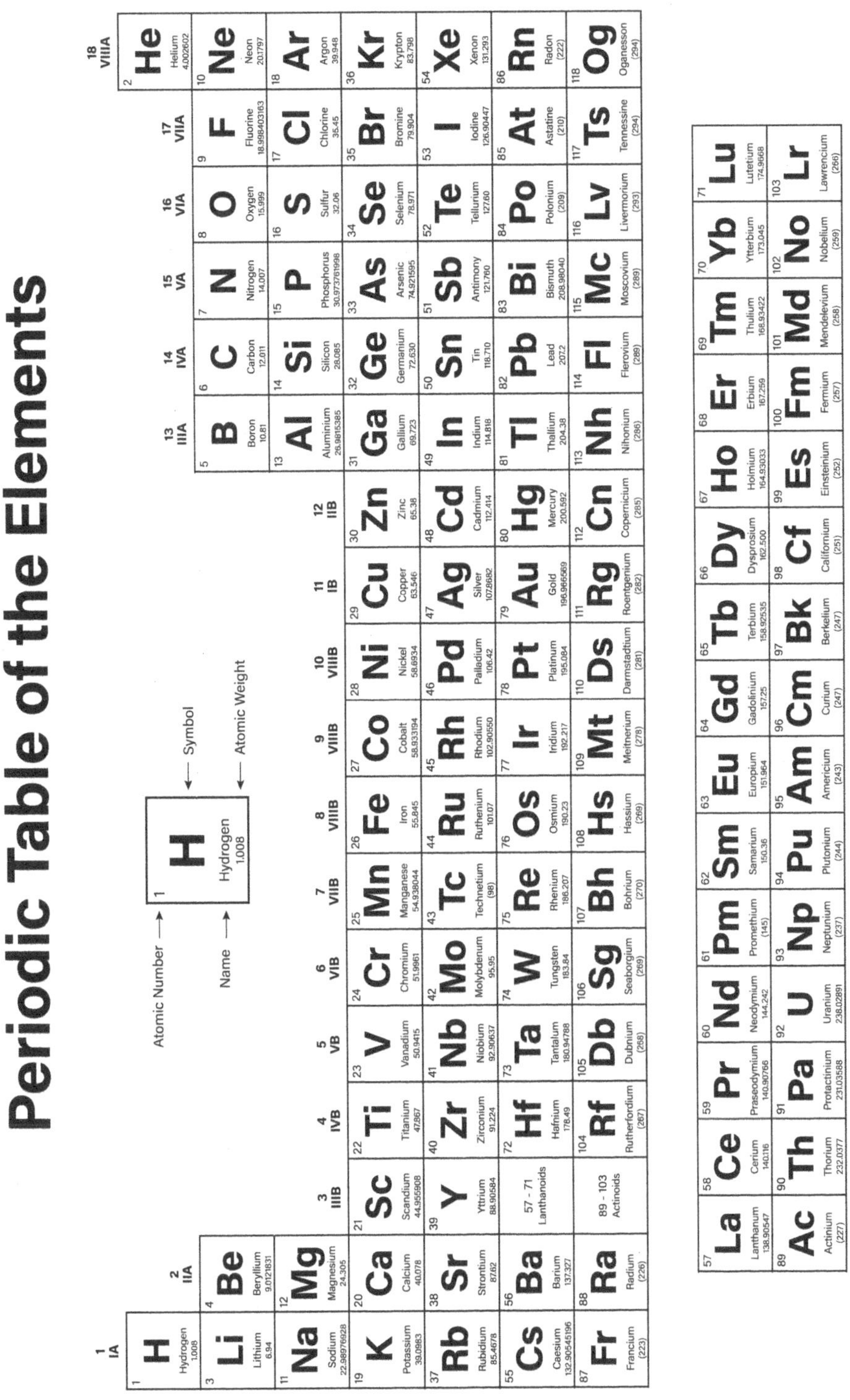

Periodic Table of the Elements

FIGURE 7.2
Periodic table.

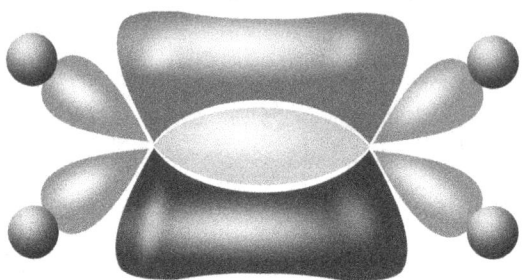

FIGURE 7.3
Hybrid orbital shape.

The first is that the energy levels that the electrons are in change to values different from the energy levels they were in before. Because color is determined by energy, the colors of the formed molecule are different from the colors of the atoms that formed them. Indeed, the diversity of available colors in the world around us comes from the fact that hybridizing orbitals change the colors to colors otherwise unseen and unknowable.

The second change is that the shapes of the orbitals hybridize in interesting and consequential fashions. Here's an example of a hybrid orbital. Notice how much more complex and pivotal it is (Figure 7.3).

If you look up the full suite of hybrids. You'll see that the s-s bonds are pretty symmetric, the s-p bonds have a lot of possible shapes, and the p-p bonds can look and act like hinges. These bonding possibilities help determine where the atoms can be relative to each other. This is usually described in terms of the angles that can form between atoms in a molecule.

The entanglement between electrons in molecules takes the probability distributions of electron positions and begins the complex process of building what we perceive as shape.

Our sense of shape comes from our perception of motion and color. The probability distributions of electrons affect how the atoms can be placed relative to each other and how they can move relative to each other and therefore what shapes the molecules can have.

How a molecule's shape can be changed depends on the entanglement of the atoms. As molecules grow in complexity, the possible positions of their atoms change based on the multiple bonds involved and the residual electromagnetic forces the atoms exert on each other.

Materials science, crystallography, and organic chemistry all explore and make use of the aspects of this growing complexity. The space of possible shapes, orientations, internal motions, and interactions grows rapidly in complexity as more atoms are added to the molecular structures.

Any student of organic chemistry has to learn a special polysynthetic vocabulary to be able to describe the individual molecules formable from Carbon (with guest stars Nitrogen, Oxygen, and Hydrogen). However, that vocabulary does not express the three-dimensional curves and branches that determine what molecules can interact with the others.

It is the four-dimensional dance of the organics that makes life possible. At each level of structure and move of the dance, more and more information is needed to accurately describe, map, and animate what is going on.

At each level of structure, probabilistic information appears from the level below, which creates possibilities of interaction at this new level, but not in simple ways. What is possible grows based on interaction and is narrowed by how things can be shaped and moved.

The information necessary to accurately describe what can happen increases greatly as complexity increases.

But, because the base information is probabilistic, it is not possible to ever describe what will definitely happen in action and interaction. In many situations, some outcomes are highly likely, but none are certain. The probabilistic character of the universe is what shapes what happens to be and what can follow from it.

And yet, we can still use the classical laws of physics in many situations if we understand them to be fuzzy descriptions of the most likely region of outcomes.

F is pretty darn close to ma where F, m, and a are all measured approximately enough for uncertainty to not mess us up.

We'll talk a great deal more about measurement and the ability to keep using classical models in Chapter 8.

Information Seen from Multiple Perspectives: Basis and Observation

There's an element of uncertainty and observation that we slipped past but need to return to. How and where observation is done determines what information is generated and what information cannot be generated.

We casually wrote $\Delta p \Delta q > h/4\pi$ if p and q are conjugate quantities, and blithely talked about the observation of one or another of these quantities.

But observation is interaction. Any act of measurement requires an interaction that produces something that our senses can perceive.

Sometimes this perception is direct, such as observing momentum by letting something hit you, or indirect, such as sticking a thermometer into something and examining the reading on the thermometer.

It is rare to be actually measuring only one of the conjugate quantities. Usually, a measure of momentum has some measurement of position about it, and a measure of energy is usually within a region of time. Most of our measurements are mixed results between fundamentally entangled values.

Our senses are based on this. We perceive position and shape using motion. We rely on these mixed methods that give us fuzzy mixed results. Whatever method we use will prevent us from using other methods to measure the same quantities because these acts of measurement change the information we would be measuring.

There are quantum information theorems about how different configurations of measuring methods determine what we can and cannot learn. The language of the theorems is that of linear algebra. The theorems show the differences in observable data using different bases in the space of information. Measuring one way gets one set of values for one set of variables and measuring another way gets a different set of values for a different incommensurable set of values.

We solve this problem of not being able to get all the needed values by doing a lot of measurements of recurrent events. We measure each event uniquely and aggregate the resulting data (yes, it's data) by records and statistics and learn about the recurrence from that. We do statistics to estimate probabilities. We find the entangled information (that is, the probability distribution) by rolling a lot of dice and tabulating and calculating from the results.

From an individual human perspective, this involves trying something over and over again to learn its ways. This can evolve scientific methodologies, technical processes, and art styles.

This metamethod helps with recurrent events. For unique events, we distribute in space rather than time by gathering information from multiple perspectives and recording multiple observations of aspects of the event. From this, we both compose aggregate pictures and make note of unique observations.

The primary human method for this is talking to a variety of people who all experienced the event and recorded their recollections.

Our learning methods of trying and conversing allow us as individuals and groups of people to solve the fundamental problem of observation by always doing things more than once either by ourselves or by sharing with our other people.

It's almost as if our methods of learning evolved in an environment where uniqueness and recurrence were aspects of nature.

Information as Part of Reality

Is information real?

That's the kind of question that tends to get Yes/No/What do you mean?/That's a philosophical question as responses.

Roughly what we're asking is, "Is information as a concept more than just a model for certain aspects of reality as humans perceive and interact with it?"

The answer to that question seems to be, "Kind of."

On the face of it, information theory seems to be an attempt to model a very human activity, the passage of knowledge from one mind to another through a medium of art/language/mathematics.

But, by the same token, probability theory was created to model the extremely human activity of determining whether or not to wager a certain amount of money or other wealth on the outcome of an event in the future, or should one instead get some food? The Earl of Sandwich answered with why not do both? – but that's a side dish matter.

The founders of probability theory would have likely been boggled to find out that their aid to gambling was also an accurate theory describing many of the fundamental underlying processes of reality.

This boggling occludes one of the basic facts of humanity, that our concepts of whether or not to take actions evolved inside a fundamentally probabilistic universe. It makes sense that the mathematics of what to try and what not to try would have to in some way model the human experience of existing in a universe of possibility where all actions are in some sense gambles.

The same is true of information theory. The models of knowledge derived from observation and possibilities evolved in a universe overflowing with interacting structures of information. At the lowest levels of this universe are fields/particles that can be characterized and distinguished by specific pieces of interacting information. At each level of the structure above, this new information is created by interactions below, and new possibilities emerge from the usages of that information.

So, just as probability permeates reality from the lowest levels passing through the human level to the highest levels of structure, so does information permeate those levels.

These two theories, probability and information, are inseparably entangled both at the quantum and human levels, spinning together in opposite but undefined directions.

Personal Exercises

1. Consider at least one category you see as sharp-edged and examine what information you rely on to put things in that category.

2. Look at the instances of category in exercise one and see how diverse the results are.

3. Examine one object and see what properties of it are inseparable.

Testable Exercises

1. After doing personal exercise 1, see if you can create a binary tree of questions to answer whether or not something belongs to that category, and then see what the other branches might refer to.

2. After doing personal exercise 2, describe how broad the space of possibilities you've found is and see which qualities that distinguish them are difficult to describe simply. Those qualities are the art.

3. After doing personal exercise 3, write down what would happen to the inseparable properties if you changed one value of one inseparable property.

8

Human Perception and Measurement

How Humans Perceive: The Filtration and Enhancement of Information

There are two philosophical extremes that have been claimed about human perception, either that we perceive reality perfectly or that all we perceive is illusion.

Mountains of books have been written using one or the other of these extremes as a foundational principle. You may have noticed that we've already talked about subjects that mountains of books have been written about, and we've tried to avoid climbing any of those mountains.

In order to keep this book to its purpose, we've been working to be a river in the valley between those mountains.

Our concern is the practicality and mentality of programming as art, not any question of essentialism. From that practical perspective, we perceive parts of reality and gain some benefit therefrom, neither perfect nor useless, but if we work at it, good enough.

Avoiding those two mountains, we can ask how human perception works, which leads us to other mountain ranges. There are piles of books about how human perception works, along with other mountains of books about the perceptions of other animals and mountains of books about what there is that could be and is perceived.

We're steering now between these mountains.

Perceptions for humans are founded on neurological processes that generate structures of neural impulses and neurotransmitters using interactions between human bodies and chunks of reality the bodies are in contact with.

The need for contact with the perceived is both important and fuzzy.

How close do things have to be to be in contact?

Close, but not as close as you might think. We'd have to dig down into quantum physics to answer this question. The short answer is close enough on the human scale for our evolved senses to get some sort of input to act on.

What input is needed depends on which sense we're talking about.

People vary in their ability to see, hear, smell, touch, taste, and other less well-known senses like proprioception. I'll be laying out fairly broad descriptions of the senses. Human diversity means that each thing below applies differently to each person.

Our senses are diverse in needed input. Sight perceives light. Light is photons. Photons are fundamental particles. Hearing perceives waves of vibration in whatever medium we are in. This is usually air, which is a mixture of gases, but sometimes we're in water and occasionally other materials. Smell and taste employ chemical receptors which are essentially checking for molecular shape. Touch is a lot of different senses. We also have

DOI: 10.1201/9781003459866-10

interior perceptions. That is, we have some awareness of our own large-scale bodily processes. Feeling hungry is an example of this, as is feeling sleepy, achy, etc. Interestingly, taste is processed in the same part of the brain as the interior perception (the insular cortex).

The interior neurological flows generated are how and what we call perception. They have relationships to what we are perceiving (the input), but they also have relationships to the sources of what we are perceiving (the generators of the output that we process as input).

Our sense of sight isn't just photosensitivity. It allows us to construct images of what the light is coming off of. Our sense of hearing associates with what's making the noise or music or speech. This is one of the most basic acts of association in human thinking.

We can look at the process of perception as having three steps.

1. What's actually interacting with our sense organs?
2. What happens in us when the interacting stuff interacts with our sense organs?
3. What associations are generated when the neurological consequences of step 2 get to the more thinky parts of our brains?

Now, we have more mountains to steer between.

We're not going to examine all the senses, but we will look at sight, hearing, and smell because they are all qualitatively different and they show three very different modes of perception.

Sight is the most digital of our senses. Our retinas consist of an array of extremely small sensors, rods, which mostly do low-light grayscale effects, and cones, which process color in brighter light, from which we assemble images. Cones are of three kinds which more or less correspond to red, green, and blue values. It's not really that simple, but it is why we think RGB images look right.

The human sense of sight is quite narrow compared to the ranges of electromagnetic radiation that are all around us. Indeed, our idea that the sky at night is mostly dark comes from the fact that we can't perceive the panoply of radio and X-ray activities happening above us. We also can't see much into the infrared and ultraviolet, which means we are lacking an awareness of how complex and brightly colored a number of plants and insects are.

If we shift from the particle model of light to the wave model, we can see that our ability to perceive light is for most people in the range of wavelengths between 380 and 700 nanometers. In musical terms, this is one octave (because the low note of an octave is twice the wavelength of its high note). Our sense of sight does not have access to the kind of harmonics that happen when waves differ by one or more octaves. We have no light equivalent to low C, middle C, and high C. We have no low green an octave below green or high violet an octave above violet.

Our eyes are also doing some weird bits of processing to get RGB and shading values for colors like orange. But again, that's deeper than we're going to examine.

The eye makes image data out of this input. What it's sending back to the thinky bits isn't the same as a bitmap made of rod and cone impulses. The optic nerve does its own processing before it hands off information to the rest of the brain.

When the information gets into the brain it is combined with recent perceptions in the process of persistence of vision. It also blends with what we have learned to recognize.

More processing happens if what we are looking at looks like a human, and especially if it looks like a human face.

These layered processes generate a fairly rapid perception of what we think we are seeing. There's a lot of work being done to create the combination of perception and recognition those of us who can see rely on.

Sight, therefore, is a highly filtered, digitized, and motion-employing sense used to give us a four-dimensional awareness of the surfaces of things in an arc in front of us that is rapidly connected to what we might be recognizing.

Hearing, by contrast, is a perception of the vibrations in the medium we happen to be immersed in or in contact with. The motion of a medium is generally modeled as waves, and waves are what we hear with our ears. Our bodies can feel beats and rhythms through our senses of touch and our internal awareness. Tones we hear in our ears.

The process of tonal perception involves surfaces, small bones, and coils of liquid that resonate with the incoming sounds. Our senses of hearing cover multiple octaves of tones so we can hear and make harmony.

But our hearing does more than just take in. The shapes of our ears add harmonics to the sounds that come in. The sounds that our auditory nerves create impulses from are more musical than the sounds that are actually being generated in the environment around us. Before our nervous systems even start processing, our bodies have already augmented the reality.

The actual generation of auditory nerve impulses is still under study. Lots of different sources of stimulation exist in the human ear (including thousands of really small hairs). So, we don't have a simple RGB correlation the way we do with sight. We do know that nerve impulses travel along the auditory nerves to the brain, which processes it to various forms of recognition. If language is recognized, then the specialized language areas kick in.

Note that the language areas are not only accessed by auditory impulses. Sign language and reading use some but not all of the same specialized areas.

Sight is digital/more or less particle-based. Sound is analog/mostly wave-based. Scent is neither.

Scent deals with the more complex phenomenon of molecular shape and bonding. Essentially, our nostrils contain a variety of shape sorters that can bond with different molecular arrangements. As we mentioned in Chapter 7, organic compounds have a wide variety of three-dimensional shapes which we can metaphorically see as a panoply of keys. Our scent receptors are like locks that check for which keys/molecules fit in them.

Scent is processed quickly and deeply and is one of the strongest memory triggers. A recognized bundle of keys can pull up a whole batch of relevant memories. This can also produce serious misidentifications. A bunch of chemical triggers does not mean you are actually smelling chicken soup.

Sight, sound, scent, and all the other senses produce rapid responses in our brains. Because of that, we often jump to immediate conclusions about what we are perceiving. We see/hear/smell and assume we know.

But our senses and brains produce a fair number of false positives and false negatives.

Misidentification of what one is perceiving can vary between humorous and lethal. This produces evolutionary pressure both to improve the quality of the humor and correct the errors in identification.

A long time ago, probably well before modern humans evolved, humans began to develop mental and physical tools and practices to slow down the process of recognition. They developed practices to be able to examine things from multiple perspectives.

These practices evolved into what we now call measurement.

Measurement and the Creation of Data

We're about to examine measurement, a concept so basic to human action that it is nearly impossible to define accurately and far too easy to create myths about what it was invented for.

For our current purposes, measurement is the process of using sensory awareness to generate data (the material of classical information theory) from ambient information (the material of quantum information theory). Data have two benefits and two drawbacks relative to information.

Benefits

1. Data can have greater exactitude than information.
2. Data are easier to convey from one person to another than information.

Drawbacks

1. Greater exactitude is not the same as greater accuracy in a universe of probability and uncertainty.
2. It is easier to convey specific aspects of what one is talking about than it is to convey complex awareness.

The primary solution to the drawbacks is to use the benefits to create structures of data that a human will be able to perceive in a way that illuminates more than a listing of the data. That solution is art.

Measurement can be seen as actions that examine specific aspects of observable phenomena and from them form communicable pieces of data that would help someone create in their own minds or out of materials something implicative of the object measured.

We are used to the idea that measurements should take the form of numbers. But metaphors can also be measurements. Wine-dark sea is a measurement of the color of the ocean. Crusty like wet sand is a measurement of texture. Funny like (comedian one really likes or dislikes) is a review as a measurement.

The effect of numbers on measurement was revolutionary. It allowed for precision in data and pointed toward the creation of tools as means of measurement and transport of measurement. My brother and I have discussed this revolution in our science popularization *Three Steps to the Universe* and our math popularization *X Marks the Spot*. So, I'm going to resist the temptation to reiterate those long discussions and instead point out two things about numerical measurement.

1. Numerical measurements can be named and isolated. The length, width, height, and mass of a box are all measurable, recordable, and transportable as data to someone else.
2. Numerical measurements can be used as parameters in calculations to produce inferences about other qualities of the measured object which can be used to predict what is likely to happen to the object if you do things to it. Length * width * height = volume which tells you how much water you should be able to pour into the object before it spills over. Mass/volume = density. If the density is less than that of water, then it should float if you drop it in water. Calculation makes inference a controllable and testable process.

Measurement of the entangled world gives us separated data from which we can infer nonmeasured aspects of the entangled world.

And herein lies the primary question about measurement. What can be measured? What data are separable from the entangled complexity of the universe?

The classic saying if all you have is a hammer all the world looks like nails applies even more so to measurement. If you are used to measurable quantities, then there is a tendency to discount the nonmeasurable as unimportant.

There is also the opposite tendency to assert that a nonmeasurable awareness is actually a measurement. We see this in discussions of taste and beauty, wherein people will take their personal reaction to something, their like or dislike, and assert that they have measured a quality of the thing.

This is so common that it is easier to make the error than not make it. If I enjoy the taste of something at the moment it is easier to say, "This is delicious" than to say, "Right now I am enjoying the taste of this."

These two statements are not even close to synonymous. The first is asserting a measurement of the object. The second is a measurement of my momentary experience of the object. Both are data, but the former is the data that generate the erroneous implication that the datum of deliciousness is universal, not local to me at this moment.

Numerical measurement is also used for comparison, which works fine for the length of nonfractal shapes, the rest mass of objects that have rest mass, and a large number of other physical quantities.

But comparison as a tool sees all pairings of objects as comparable. Thus, it creates a tendency to compare the noncomparable. From this, insidious questions arise, such as "Which of your children do you love most?"

Measurement and comparison also create a tendency to eternalize transitory feelings. "What's your favorite food?" largely depends on what one feels like eating at the moment. "What's your favorite color?" has even broader problems because it has no context whatsoever.

Measurement also carries the recently discovered problems of uncertainty and entanglement. You can't take all the measurements of something. Every measurement you take will preclude or at least restrict the accuracy of other measurements you might wish to take.

Measurements need to be taken with care and used with care.

But how do you take those measurements? What tools do you have? What tools should you use? How accurate do your measurements need to be? Do you need to take the measurements yourself or can you rely on someone else's measurements and the accuracy of their recording of the measurements and the honesty of their reporting to you of the measurements they took?

These questions can seem paralyzing if asked in the abstract as I've just done. However, in any real-world situation when you are taking measurements, the answers will come pretty quickly. As we'll see in Chapter 14, the needs of the situation will provide answers or clues to how to answer the questions. It is only in this abstract place where we are looking at measurement as a concept that the questions can seem overwhelming.

Measurement is fundamentally a practical matter. It is about deriving useful data from the ambient information. The combination of what you are measuring and what you are going to do with that information will tell you what kind of measurements and measuring tools you will need.

Often the more complicated questions are about what perspectives you should take measurements from, what scale of measurement you should work from, how fast you should be moving relative to what you're measuring, and how fuzzy your results can be and still be usable.

These are the questions that cause observatories to be built on tall mountains, photographers to move too close to active volcanoes, artists' models to stand almost unmoving for hours on end, and a host of other complications arising from the fact that reality does not in fact stay still.

The Artist's Eye

Every art requires an ability to examine the objects and processes in reality that the art depicts in a way that helps the artist figure out how to depict it.

This skill develops over time and never finishes. The more practice one has with perceiving the world relative to one's art, the more things one will perceive relative to that art which open up more possibilities for the art.

All of the chapters in this section contain considerations and practices that might help develop this skill in relation to the art of programming.

Here we're going to take a look at the skill as it applies to all arts. Its basic tools are measurement and the use of measured data to determine what materials and techniques to apply in order to create an illuminating artwork from the aspects of reality one was measuring.

We're going to use mostly visual metaphors and examples for this skill (there will also be some cooking). Because of this stricture, we'll call this skill the artist's eye, but it can also be an artist's ear or touch or mouth or nose.

The choice of artist's eye as a term is a matter of my personal limitations. I have not had enough practice with this skill in nonvisual and nonculinary arts to be able to convey this as well as I should.

This is an inherent problem in writing a book like this. The author or authors will always work from what they understand of art outward and will have the easiest time elaborating what they know how to do. My perspectives are present in the naming convention I'm using. Change it as you need to, and remember that everything I'm saying has my perspective in it. This too is a truth of art.

Two questions arise in parallel when examining an object for use in art.

1. What are you measuring with?
2. What are you measuring for?

You may not initially know the answer to the second question. Something you've seen or heard or smelled or touched may simply be intriguing at the moment. You may have no coherent idea yet of what you are going to do with it.

But even in that situation you have a first and vital answer to the first question. You are measuring with your mind and senses. Everything else you might use to answer the first question is a tool you are adjoining as an aid to your mind and senses. You are measuring what looks important to you right now.

Move around the thing and find what looks intriguing. If you have generally useful tools like cameras or sketch books or audio recorders or notebooks, use them.

Note: The above advice presumes that you have appropriate permission to move about, record, etc. If you are in someone else's space, do not be intrusive. Artists can develop an unfortunate idea that the whole world is theirs to make use of. Be wary of this.

If you know what you are measuring for, you will likely have an easier time knowing what measurements to take. You may also know how to record the information for your later use, but maybe not. You might look at a very complex object you'd like to draw or paint or sculpt or make a 3D model of and you can't see immediately what aspects of it make it work for you. In that situation, it's good to take views from multiple directions.

If you're listening to someone talk in order to base a written or acted character on how they talk, it can be tricky to find the crucial elements. In this case, recording is very valuable (if they're okay with it).

Some arts, such as cooking, remain idiosyncratic in terms of what to measure. A cook trying to adapt a dish they've eaten will likely taste and look relative to their personal methods. Don't be afraid to adapt what you're tasting to what you enjoy. You don't need to be a good copyist to be a good cook.

There is a crucial part of developing an artist's eye that many people don't notice needs doing, the removal of recognition and objectification.

Perception evolved along with recognition, the ability to rapidly know what you're looking at.

But, just as measurement requires slowing down, the process of generating sense data and examining things from multiple points of view with appropriate tools, so recognition needs to slow down and learn to not know what it's looking at.

The reason for this is simple. What we know, we project. We make our augmented reality using the entanglement of perception and knowledge.

But if we are looking at something in order to depict it, we need to perceive it without knowing. To do that, it is often necessary to actively take away our presumptions.

This is at its most important when looking at something one has biases about. The history of visual art is replete with paintings, drawings, sculptures, and comic strips wherein some of the people depicted were clearly seen with an artist's eye and others are simply reiterations of the prejudices of the people and culture of the time and place. The artists did not know they were doing this. It's clear from the compositions of their works that the carefully depicted and the stereotyped were seen as both having the same level of accuracy.

A good test to see if one might be in this situation is to check how certain one is about what one is looking at or hearing. The more certainty, the less accuracy. Look, listen, smell, and examine from different angles and scales until certainty falls away, and then begin again.

Look at a salt shaker until you can't see a tool for shaking salt, but instead, you are perceiving its shape and materials. Listen to music until you can't hear the composition, but you can hear the actions and sounds of each instrument. Look at clothes until you can't see the fashion of them, only the material, hang, and flow.

Don't worry if you can't actually do these practices in their full form. What matters is the separation of perception and recognition. The ways that work for you to do this can be unique to you and may require a good deal of experimentation before you find them.

Human brains have a lot of processing power dedicated to parsing, recognizing, and interacting with other humans. We measure a lot more about people we pay attention to than we do anything else in reality. Humans are unique and capable of recognizing each other's uniqueness. Uniqueness appears when we stop being certain of who we're looking at because then we pay attention to more factors with more dedicated processing power.

Persistence and the Ability to Accumulate Awareness: Making New Information from Data

Having broken down our own recognition in order to be able to see what's in front of us, measure it, and gather data, we must now move on to the task of taking that data and from it producing something that other people will be able to make use of. That is, we wish to make something that some other people will be able to recognize, gather data from, and form information from it in their own minds that will help them in their work, play, learning, and/or sharing.

In short, having observed something in our augmented reality, we wish to use our observations to add new objects and awareness to other people's augmented reality.

Everything humans do is art because we perceive art and use that to make more art for others to perceive.

To do this, we rely on three things we know about our human audience:

1. Human perception has the ability to create false positives in recognition.

2. A human mind has the ability to connect what it recognizes with what it knows.

3. The art we make will have an extent and a persistence in time.

Our art has to be large enough to be perceived and last long enough for humans to notice it and make use of it.

Until recently, it was very hard to make things that did not fit human scales of space and time.

We programmers, however, practice one of the arts where the actions involved happen much faster than humans can perceive and happen buried inside boxes that don't have to show anything.

For the human interaction side, we need to pick and display the data the user will need on human timescales and human-perceivable sizes. It's why input and output are techniques for us.

Whatever the art, the data being presented to audiences need to be perceivable, composable, and contrastable.

A painter will use multiple paints to create a recognizable image, relying on the way the human eye composes perceived colors into something like an RGB image. The paints give hints to the mind, which then creates recognition from it.

Animators present images in succession that are similar enough that the mind will interpolate the idea that they are the same objects and beings, but different enough to give a sense that some of those objects are in motion.

A cook will combine ingredients that are chemically tasteable in such a way as to give the mind the sense that what is being tasted harmonizes.

All of these will work for some people some of the time. That's the basic truth of artistic recognition. Some minds will compose the persistent artistic data and from it grow more aware in some ways, and other minds won't.

In programming, we have an interesting tool for visualization that was first developed by mathematicians, the act of making graphs.

In broad terms, graphing is the ability to present abstract information in shapes in a way that most human minds would see as coherent. The graph maintains the abstraction of the data (e.g., time increases as one moves to the right and the number of potato chips eaten increases as one moves upward) but uses the recognition of shape to give some sense of the relationship between the left–right dimension and the up–down dimension.

With computers, we can expand this to a ten-dimensional structure, each dimension of which can be meaningful and visible in different ways.

The ten dimensions are three for space, three for velocity, one for time, and three for color. We can present meaningful evolving graphs on monitors by using up to ten measured values as parameters for the shapes being presented.

It's a bit tricky to make this work and is in fact better if the ten dimensions are not all independent.

We can push it to 11 if we add transparency to our colors.

We can push it to a great many more dimensions if we encode evolving data as music.

Here is where things get subtle and arty.

The ability to represent a lot of evolving data in a variety of fashions requires that the data compose together in a way that the perceiving mind can make use of as it makes use of things. If your presentation is redundant such that the same data are visible and/or audible in a variety of ways, then your audience will be wider then, if you insist that your users be able to decode the ten-dimensional presentation you are making.

Workable and enjoyable art has hidden redundancies and harmonies. It's fuzzy and tangly and has implications for how it looks and moves. It is easy to gather in and turn around and see from multiple sides. Ideally, it should be possible for many people to all say, "Oh, I get it," and each of them is getting what they personally need and/or enjoy from what they are looking at. If they later talk about it, then they may find that each of them got something different from the art and, in discussion, they share these different views and can put together a wider understanding than anyone including the artist had of the artwork.

Presence and the Biasing of Awareness

What is present in the audience's mind can greatly affect what new art they perceive. Just as an artist's preconceptions and certainties can get in the way of the art they are making, so can the audience's preconceptions and certainties interfere with their ability to benefit from art that does not fit those certainties.

Artists need to be aware of such presences but need not necessarily tailor their art relative to them. Many writers have tried to tiptoe around prejudices and social conventions, only to discover that this only strengthens the presence they are trying to avoid getting near. The presence is called by trying to avoid it.

It is generally wiser to make your art honestly and clearly and show what you are seeking to show in as multiply perceivable ways as possible.

This is not to say that presences are inherently bad. All art relies upon and generates a presence of how it moves. Making guidance for those motions to be a part of the art itself can be of great benefit to the audience.

For example, many dances in the 20th century were introduced with songs that contained lyrics saying how to do the dance set to music that the dance could be danced to. The dance, the song, the instructions, and the music were all presented in one bundled presence.

Programming can do this more easily because we can construct our own interfaces and embed the presence of the program in its shape and action.

We need to be cautious that such a presence not be a burden to users it doesn't work for. Users should, as we've said, be able to do things in multiple ways. If the user is aware of the ability to make the experience that works for them, then that ability to adapt will be part of the presence of the program to them.

Presences in art and people's minds can themselves compose and contrast. This is idiosyncratic. The artworks that speak to and inspire a given person will always be an eclectic collection entangled through personal experiences and the threads of their lives.

It's good to be aware that any art you make might end up mattering a great deal to someone in ways you could not anticipate. One of the joys of art making is finding out that it sets someone going in a way that neither you nor they thought possible.

Presences can also find their way into the conversations about art and a culture's awareness of what the art depicts. This is often called intertextuality, the awareness and knowledge that socially lies between artworks.

Symmetry Making and Breaking

A number of laws of physics have inherent symmetries. Newton's third law of motion says that to every action there is an equal and opposite reaction. His law of universal gravitation says that the gravitational force exerted on an object of mass $m1$ by another object of mass $m2$ is $Gm1m2/r^2$, where r is the distance between the objects. So, the forces operating on both masses are equal.

There is a theorem famously proven by mathematician Emmy Noether that any conserved quantity has a related (often hidden) symmetry.

Symmetries reduce the amount of data needed to determine things. A circle is defined as the set of all points in a plane a given distance (the radius) from a given point (the center). The same definition in space gives a sphere. So, the whole set of points making up a circle or sphere (or higher-dimensional sphere) is defined by the symmetry of having the same distance around a particular point. Therefore, the ability to make a circle or sphere requires only the data of the center and radius. The act of drawing the circle requires the ability to use that data to generate those points (Figure 8.1).

Some symmetries are more complex than these examples. The theory of relativity gives a way to transform any position and momentum data measured from one frame of reference to what would be measured from another frame of reference. This involves some pretty messy calculations, but it can be done.

Symmetry produces reversible operations. Rotate 45 degrees to the right, then 45 degrees to the left, and you're back to the same perspective as you started with. The same applies if you walk five meters forward and five meters back.

There is a beauty in symmetry.

But what causes evolution, change, growth, and transformation of possibility is breaking symmetries. The early universe had some fascinating broken symmetries, including the Higgs mechanism by which matter acquired mass.

Two entangled electrons represent a symmetry. The electrons are interchangeable as long as they are not observed. Once observed, a metaphorical coin is flipped and one of them is spin up and the other spin down. A bit has been set that was not set before.

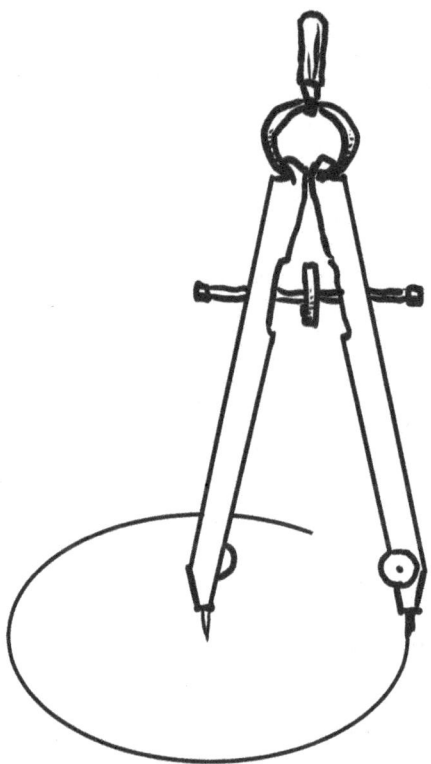

FIGURE 8.1
A compass with a partially drawn circle.

Entangle two electrons in the outer shells of two atoms and you have made a symmetry in the bond but altered the motions and possibilities of the atoms. Making the symmetry on the electron level creates asymmetries on the atomic and molecular levels.

The reason I bring this up is that the processes of making and breaking symmetries are fundamental to artmaking. Determining whether the data you are placing in your art is best embodied as symmetry or broken symmetry is a recurrent artistic consideration.

How alike things are shaped, how harmonious their colors and sounds, how predictable a plot is, how chords progress in music, how smooth a fabric is, and how similar the left and right sides of a garment are all matters of embodiment in symmetry and broken symmetry.

If your art depicts events in time, how reversible are the actions you are having occur? As mentioned before, the ability to load saved files gives programming the ability to reverse time. The Undo menu entry adds a symmetry to time that does not exist elsewhere in reality. But as also mentioned, the undo does not change the knowledge of the user so they can learn from the error they made. Undo is symmetry within the data and broken symmetry in the user's information.

Many computer roleplaying games are built on the assumption that the character the user is playing will die repeatedly, while the user learns to navigate a puzzle or find the weak spot of a boss monster. Again, the symmetry in the game is predicated on the broken symmetry of the user's learning.

There are fundamental broken symmetries in the universe. Because of this, there are arrows of time that cannot be reversed. One of those arrows is entropy. In any closed system, such as the universe, entropy will always increase in the direction of increase in time.

Entropy is often mischaracterized as a slide toward hopeless heat death. But what entropy means in both thermodynamics and information theory is that possible configurations increase over time. More possibility means more entangled information. More entangled information means more measurables. More measurables means more things art can be made from and about.

Entropy always increasing means that as long as there are artists and audience, there will always be spaces for new art and new experiences of art. Because of entropy, art is inexhaustible.

Broken Symmetries and Tests: Questions Embodied

In Chapter 7, we contrasted a broken symmetry between the idea of recurrent phenomena, such as Gnosticism, and individual things arising within one version of that recurrent phenomenon, such as Chinese Manichean paintings. Here, we're going to go backward and create a symmetry out of that broken symmetry by examining questions and tests.

The branching tree structure with fixed questions and one-bit answers that we outlined omits a critical step: The actual calculated test. How does one determine which way one has gone down the tree? The tree shows all possible paths but does not ask how you know which path you've taken.

Tests are done by measuring and calculating. The tree is a bundle of symmetries. Any particular answer is gotten to by broken symmetries and those symmetries will have been broken in many ways.

Couliano's *Tree of Gnosis* does not ask how any one particular form of Gnosticism got the way it was. He was proposing recurrence as a phenomenon in history where classically history has focused on inheritance, on the paths of the particular. He was illuminating the idea of recurrence and a tool thereof. He was displaying a map of the geography of Gnosticism, rather than answering the geologic question of how that geography came about.

The symmetry of recurrence and the broken symmetry of inheritance come together in paired questions.

1. What characteristics does this branch have?

2. How did each thing that is in this branch get these characteristics?

The first question is answerable with data. All it takes is a batch of bits, one for each branching.

The second requires information, deep entangled information, because actual unfoldings of history involve lots of people doing lots of things at lots of times in lots of ways and their actions affect each other.

Both of these questions and both of these answers illuminate what is going on. The what question can be calculated. The how question must be dug into. Both are kinds of tests. One can be done by computer alone, the other requires human aggregation and thought.

Art should leave room for both. Persistence should evolve from both the calculated and the considered. Each should help the other.

Design your software with this pairing of human and computer in mind. Code so that the humans can set the computers to do the kind of tests they do well and the computers can produce results that can help humans contextualize their own ideas and consider possibilities of awareness and recognition.

Here's the critical thing about the entangling of art and science. Every human is part of the evolving awareness of possibility. We're all playing with it whether we know it or not. As we do this, history evolves in recurrent action and unique art.

If your art illuminates and interests, then it will be part of someone's awareness. The art will become a part of seeing and imagining questions and possibilities that might not have been thought of before because the possibilities weren't practical before.

1. When symmetries are put together, questions form.

2. When symmetries are broken, answers form.

3. Then new symmetries can form.

   - unless there are missing possibilities. And alternative meanings to the questions and answers that people have been playing with.

   - unless the tests given don't fit the facts or the needs and need remaking.

   - unless the underlying assumptions and the overt presentation don't jibe with what's actually going on in the lives of the people using the tools and experiencing the art.

If you record the results of tests and compare them to reality, you'll be able to discover if the tests themselves need remaking. If they give too many false positives or false negatives, or if they mismatch what and how, then the tests and the analysis of the tests by metatests can lead to a new iteration of the making and breaking of symmetries.

Because there is no last step, no final question, and no definitive answer. As possibility and awareness of possibility grows, the possibility of new and clearer views grows.

Possibility grows because entropy is a process of feedback and iteration.

The Surprises of Iteration

Here is the basic loop of that process:

1. Measure the information in reality to produce data.

2. Test, compose, and contrast the data to produce art.

3. Embed the art in reality.

4. Go back to step 1.

There are two forms of this loop.

The first, where humans do it, is the evolution of art in human life.

The second, where the loop is automated, happens in nature and computers.

In both of these cases, the iteration of measure, build, and embed produces objects and processes that follow the mathematics of iterated functions and the mathematics of fractals and chaos.

These are fields of study that show how simple it is to create irreversible broken symmetry from a seemingly symmetric process. Do the same thing over and over again. What could be more symmetric?

But, of course, you're doing the same thing to different things each time because you're applying the function to the output of the previous iteration of the same function.

It is a curious artifact of mathematical history that people actually thought there were iterative functions that wouldn't produce hidden chaotic explosions and fractal growths.

The simplest iterative process of adding one seems straightforward. Keep stepping along one pace at a time. But this process has hidden depths if you look at the numbers produced not in terms of the addition of one over and over, but in terms of their multiplicative prime factors.

$$1, 2, 3, 2^2, 5, 2*3, 7, 2^3, 3^2, 2*5, 11, 2^2*3, 13 \ldots$$

Keep adding one and you will find yourself bouncing all through the mappable-but-as-yet-unsolvable space of the distribution of primes and the evolving depths of number theory.

The art of programming is the art of evolving persistence. Computers are very good at the persistence and the calculation of what might come next. In using computers to do this, we find ourselves back in the first problem of data. How good is good enough?

It takes an infinite amount of data to represent a single real number.

We got around that by pointing out that any individual calculation won't need too many significant digits.

We didn't know when we made that approximation in the early days of computing that the mathematics of chaos would demand arbitrary amounts of significance simply by iterating over and over again. No matter how accurately one measures a chaotic process like weather, for example, it doesn't take long (about a week in the case of weather) for one's measurements to no longer be useful predictors.

To predict the chaos, we would need to measure more accurately than the universe and its uncertainty will let us.

Perhaps, we are looking at data from the wrong direction.

Personal Exercises

1. Take a few minutes for each sense you have and examine what input you are receiving and how you associate it with what you think its sources are.

2. Take an inventory of how many measuring tools you personally own and how many sources of measurement you rely on in daily life.

3. Do whichever of the practices outlined at the end of the Artist's Eye section you can do.

Testable Exercises

1. Make note of things you misidentify and how you go about correcting the error. Write down at least five instances.

2. Take three measuring tools you use. Write down what they measure and what requirements there are for what you are measuring to be measured by them. Note how much information needs to be within acceptable ranges before you can generate a single piece of data.

3. After doing personal exercise 3, write down any ideas you may have had for art involving any of the things or people you were examining.

9

Data as Information at a Distance

Too Much Information

Consider the stars. At our vast distances from them, we could for millennia gather only small bits of data about them, such as luminosity, color, and apparent path across the sky. We did not even know that our sun is a star. Our sun, seen at a great but not as vast a distance, had other data we could gather, such as its apparent size, the spectrum of light emitted, and its warming qualities.

When we could train telescopes on the sun, we discovered that it had shapes on it, sunspots, solar flares, and more. The more we studied the sun, the more complex it became and the more our bits of data became regions of evolving entangled information.

Using the awareness that stars are suns, we realized that they shared that complexity.

Our perceptions, our imaginations, and all of our thinking evolved in information-rich environments – too information rich. At nearly every scale we can perceive and most of the scales we've made tools to help us perceive, we find a world full of complex interactive information.

We think of measurement as giving us tools of exactness that will give us clearer awareness of what we are perceiving. Measurement lets us focus in on aspects of the information world and extract them.

The basic act of counting lets us go from the vague phrase "herd of cattle" to "twenty-seven cows, four bulls, and three calves." That's focusing on.

It's focusing one step in after we took two steps back. One herd of cattle is a two-step back measurement that produced a single piece of data. We took two steps back and then one step forward to go from seeing cattle in the field to the single datum of one herd to the three pieces of data: 27 cows, four bulls, and three calves.

The full, dense information shows up when we look at each of the animals and see how each one looks, moves, lives, interacts, what and where they eat, what their personalities are like, and so on.

Everything in our lives is like that. No matter what we're perceiving, there's too much information. Every action taken and every interaction that happens affects a great many things and has implications in more directions than we can count because the number changes moment by moment. In reality, we cannot change only one bit of data. The entanglement of information and the fundamental interactivity of the universe makes it impossible to completely isolate anything from other things.

But we can do enough separation to let us attend to a few crucial facts and changes at a time. Our filtering senses help us take in small enough amounts of data to get on with.

DOI: 10.1201/9781003459866-11

They let us attend to what is momentarily relevant and consider what we can do with those small parts of interacting, entangled reality.

The physical and mental acts of focus consist of stepping back from the awareness of something within its environment to see an isolated object and then stepping in again to see the characteristics of the now-isolated object.

This two-back-one-forward is the precursor to measurement. It's how we get data from information.

From a programming perspective, human input capabilities are impressive in their ability to change what they are inputting from. We can focus our senses on regions of the environment that our senses have previously perceived. See the cow in the field, focus on the cow. Hear the music, listen for the strings. Smell the soup in the air and follow it to the kitchen. Wrap up in the blanket and feel the fuzziness.

We can also perceive the panorama of life going on around us and move through it without specifying focus.

And we have a pretty good alert system that can learn to hear crying, smell smoke, see oncoming traffic, and quickly refocus on those areas of concern.

Human output capabilities are not as broad-reaching as input. We can perceive what we cannot affect (e.g., stars). Though we cannot touch the stars, we can map them and plot their future courses. We can also assemble what we've taken in ways that are startlingly unlikely.

See the blue shininess of these rocks. Smash the rocks, separate the shiny bits, and grind them to a fine powder. Store that and set it aside.

Separate yolks of chicken eggs. Whisk them lightly. Mix in the fine blue powder. You now have blue egg tempera paint.

Grow flax plants (which is itself a whole complicated process). Harvest them. Separate the flax filaments. Spin them together to make threads. Weave the threads into approximate rectangles. Do a lot more work and you have a canvas.

Take hairs from a Kolinsky sable. Attach the hairs in a bunch to a smoothed stick of wood with glue (how you make the glue is a lot of other functions done in succession).

You now have paint, canvas, and brush. Hand those to a person who has practiced painting, and they can focus on something they can see, remember, imagine, and paint a picture of it recognizable to some other humans.

This process can be organized in such a way that each part of it is handled by different groups of people so that one person can get the rocks, another can grind them up, another can mix the paint, and so on.

At each step of the process, the humans involved are operating by focusing input and output within the too-much-information environment. They are disassembling, transforming, and assembling pieces of reality to produce phenomenally unlikely results.

The too-much-information environment produces strong environmental pressures for the ability to focus, the ability to shift focus, and the ability to do things with what is focused on. Different organisms have evolved different configurations of responses to these pressures. What we do looks impressive and useful to us, but so does what lots of other organisms do.

All senses can be seen as extracting data regardless of what creature is doing the sensing. Most, if not all creatures, are doing the same two steps back, one step forward that we do to focus and extract. All creatures have their own output methods for what to do with the data and the real objects the data arise from. Many creatures diversify and share the results of multiple input/output tasks.

We presume that our thinking must be qualitatively different from that of other creatures, in part because of how well we make and use tools, both physical and mental, but we don't understand the minds of other creatures well enough to come to that conclusion.

Too Little Data

We need to see data in two different lights, the measurement of something and the result of the measurement. We often blur the distinction.

The length of a table is a measurable quantity. 2.6 meters is a result of the measurement.

"This table is 2.6 meters long" is a statement that connects both to the table and to the result of measuring the table.

If we use that measurement to figure out if we can get the table up a flight of stairs and through a door, we are using both connections.

If, on the other hand, we are gathering data on table sizes for a statistical analysis, then we are separating the datum of table length from the table itself.

And if, on the third hand, we chop off a chunk of the table to get firewood, then the datum "length of the table" is no longer the value of 2.6 meters.

We have distinct uses for the data as just data and the data as an implication of information. There is a benefit in being able to use either or both meanings, but we do need to be sure we know what we're doing.

There is one more complication in the meaning of data. Is the result of a measurement data if it is not recorded or made use of? Does it remain data if all recordings of it are removed, falsified, or replaced? Is it data to me if I can't access it?

These questions are not entirely abstract. Just as the too much information we exist in is relative to where we are and what we perceive, the data we use is relative to what data we have access to and what data we can make use of.

Parsing the meaning of data requires the ability to connect it back to what it is the measure of and what the consequences of that are.

A table of masses, charges, positions, and velocities of a set of objects is informative to a person who understands how gravity and electromagnetic forces work. Presented with such a table, such a person would likely infer that the data have been gathered to figure out the evolving motion of the bodies because that is a set of data needed to do such a calculation.

A list of names and addresses is less informative as to purpose and could be equally at home as a mailing list or a clue to an assassin's targets in a thriller novel.

Both of these usages are based on the awareness that addresses can be used to reach people for any of a broad range of purposes.

Without clues marked upon or within the data, we cannot know what to do with what is available to us.

This leads to more complications. Data may be labeled but the labels may not be accurate. A listing of people's names and the amount of money they supposedly have in various banks doesn't need to be true. Someone could have made the information up. Claims need to be tested and those tests cannot just be referring back to the data that make up the claim.

The data environment we exist in is sparse compared to the information environment because data are extracted and stored separately where information is bubbling up all around us. Each datum is a distinct measurement or a specific result of a calculation. If the information environment resembles the Earth around us with visible and invisible interconnections and dependencies, the data environment looks like the sky with only the visible stars.

Humans looking at sparse environments like the night sky draw connections between the data they observe.

Just as the constellations are our projection of drawn connections between stars that do not inherently have much to do with each other, so too our uses of data are drawn by bringing data together and making our own inferences and constructions from them.

We make constellations out of shapes of things that we know or imagine. We project our own awareness and usage upon the data, shaping them by our understanding.

The sparse separation of data gives us room to play with what we have abstracted. We can form shapes that are as wholly unrelated to the sources of the data as the painting of a human figure has to do with the rocks, eggs, flax, animal hairs, and sticks that it is made from.

Composition and Contrast of Data

In programming, we need to arrange data in ways that are useful for what we're doing with them.

This presents three different arrangement needs: The needs of the software, the needs of the programmer, and the needs of the user.

Software should be able to access and employ the data it is going to make use of with no more complications than necessary. These are pretty much standard efficiency and security considerations. If critical data are scattered across multiple countries and need to be accessed by multiple accesses, then perhaps you should consolidate the data and store it together, at least temporarily. If you only need a few pieces of data out of masses of the stuff, create quick access methods. If you don't know which small pieces, use lazy loading, and so on.

What a programmer needs in terms of data composition is the ease in mental associations, so that their res of the software can more easily create how the data will interact with the dictum and can more easily distinguish between pieces of data that might otherwise be confused.

There are a number of ways to go about this. None of them are right or wrong. It really depends on what works for the specific programmer.

Elaborating on an earlier example, suppose you have data on a group of objects with position, velocity, mass, and charge for each object, and you are going to write software to evolve the motion of this system in multiple steps over time.

One possible underlying calculation involves figuring out the force vectors acting on each object by calculating the gravitational force and electric force on each object, which are easy for a computer to calculate given this data. You might, if you are being more exact, also calculate magnetic forces, which can be approximated given the data. Assuming you are dealing with macroscopic objects moving at nonrelativistic speeds, you then use $F = ma$ for each object and move everything under the resulting forces for a short period of time to calculate new positions and velocities.

Assume you need to keep a stored record of each state of the evolution for animation or analytic purposes.

Here are the three different ways to arrange the data composition and contrast.

1. Before you do a step, you store the data for each object in a file that has the time represented as a part of the file's metadata. You then do the motion calculations and update each object so that it has the new information. You are now ready to go again.

2. After you do a step, you create a new data object to represent the data for a particular object in motion at the new time. You aggregate the different data values for each moving object as a time-indexed array of data elements. This makes it easy to analyze and draw the evolving path of each object.

3. After you do a step, you create a data structure with the current information for all the objects. This gives a snapshot of current conditions, which makes it easy to analyze relative positions and motions evolving over time and easy to present pictures of each moment in time.

There are calculational advantages and disadvantages to each method. What matters is what works for the programmer. Each method has a different focus and presents a different awareness of the same information.

This by no means exhausts the possibilities, but presenting three qualitatively different alternatives tends to get minds and imaginations working. There's a reason why writers often present possibilities in threes. And yes, that is a demonstration of how data composition and contrast can be used in other arts.

The main thing is to employ what works best for the way the programmer imagines the program going and how the data are best visible to the programmer.

Data as Presence and Evolving Persistence

What a user needs from the composition and contrast of data is clarity, illumination, and the ability to get a hold of what is happening and interact with it.

The user's needs depend on what information they will be connecting with the data. Their attention is given to data that has relevance to the aspect of their life and actions. What the user uses is the data composed into a persistent presence that they can incorporate into the too much information they live in.

How the data are presented determines how easily the user can use them. Each form of presentation will be of use to a different set of users.

Continuing the example above, the user can be presented with the raw data of each object in the evolution as a batch of numbers. This can be useful for people skilled in internalizing and analyzing this kind of data and looking for anomalies.

This would be useful if, for example, the data were taken from reality and the actual motions needed to be compared to the predicted motions. In this case, the data would need to be output in a format similar to that of the recorded data of the objects in motion so that comparisons could easily be calculated. One might do this if one were looking for the presence of other forces in action (e.g., from nonvisible objects).

The data might be presented as a set of paths drawn in space, showing the ways each object traverses over the course of the time evolution. This would give an overall view of the whole process. If the paths were drawn with a gradient color, the user could get a sense of where each object is at a given time.

Alternatively, a frame-by-frame animation could be created to show the evolution as it unfolds. Such a view could be seen from either an outside perspective or calculated from a POV of one of the objects.

The overall-view animation could be combined with the paths of the previous perspective to show both individual and group evolution. The numbers could also be placed with each moving object.

At this point, the user could experience more data than they can easily parse. It's all too easy to recreate too much information by dumping a lot of data in a small space.

You can give users the option to turn on or off the different aspects and to control the speed of the animation; so, they can glean what they need to.

It is important to be aware that too much is too much. Many people find it easier to turn options on starting from a simple view and then to turn options off from a complex view.

We tend to try to make sense of what is shown us. Taking something away for ease of awareness often feels like one is not dealing with what's actually happening. Adding something feels like one has gained more of what there is to understand.

4 − 1 = 3 by taking away.

2 + 1 = 3 by adding.

Both get one to 3 but the path matters for how one feels about it.

It's a curious bit of psychology because the whole process of measurement and data gathering arises from exactly such a process of taking away until what remains is easy to measure. Taking apart and then putting together is a two-steps-back-one-forward that can feel like losing the way and then finding it again even though both forward and back are steps along the way.

Art Is Made from Data through the Action of Changing Information

One of the benefits of data to artmaking is the ability to blend in order to make compositions and to create gaps to make contrasts. The artist can use data mixing and separating methods to produce artificially smooth combinations and artificially sharp contrasts.

Reality tends toward a fractal complexity with a concomitant complexity of entangled information.

Smooth and sharp are both distant perspectives, the kind of artificial focus that produces and uses data.

Human perception at a distance can create the impression of sharp lines and smooth blends. If data are measured at such a distance, artistic contrasts and combinations of data can replicate the impression of seeing information at a distance. The distance can also be used to create different impressions entirely by composing different awarenesses and perspectives applied to the extracted data.

Because of this, it is possible to use data that were drawn from one region of information to convey an awareness of a completely unconnected region of information. The reality from which we get paint rarely has anything to do with what it is painted on or what it is painted to show.

This kind of transformation of data from meaningful to meaningless to differently meaningful is one of the most intriguing parts of art and human psychology. We form recognition in a way that generates many false positives because we don't insist that something must be what it came from.

Tyrian purple (a hot pink dye made from a species of sea snail) is not associated with the snails or the sea or even Tyre. Its most famous association is as a color worn by Roman emperors.

The sideways association with connection and disconnection, lost and gained meaning, is what allows art to be flexible and changeable. Materials become materials by extracting and combining them. Techniques become techniques by practicing them. Tools become tools by refining and diversifying them.

If you look at the sparse data you are working with as balanced upon your fingertips waiting to go where you place it, you can pick up pieces of earth and images of sky and assemble them in ways that work through how you work. And once given over to an audience can be used by them to fit into their personal needs and enjoyments.

Personal Exercises

1. Consider your personal work- and play-spaces. Consider how you have objects arranged there. How do you make use of the proximity and separation between them?

2. Practice the two-steps back, one forward method on a variety of sources of sights and sounds. Move far enough back so that things blend together, then step forward again until they acquire some distinctions, but not so close that they are full of details.

Testable Exercises

1. Document at least one use of personal exercise 1.

2. Take one object you own. Research in not too much detail what the components that make it up are, and if possible, find out where they came from and how many different people were involved in making it.

3. Make note of various forms of graphing you have seen used to present data. Document which ones are most communicative to you and which confound or confuse you.

10

Data Structures as Architecture and Furniture for Data

One of the critical, but not often thought of considerations in designing data structures is how easily your thinking can flow through them. If every time you need to work with a data structure you need to remind yourself how it represents what it represents, you may need to redesign it, so it works better with how you think.

Before we dive into this, let's take a step back (or two back, one forward) and consider the mathematical concept underlying all data structures: The variable.

A variable is a symbol that can range over all the elements in a specified set. If x is a real number, then x can take any value from the real numbers and we can employ it in any calculation, test, or data structure that employs or operates on real numbers.

The set that the variable ranges over is expressed in terms of a diverse collection of objects that share a characteristic in common. So, x is any (diversity) real number (common characteristic).

Data structuring arises from looking inside the commonality for ways to express diversity.

If p represents any point in Euclidean two-dimensional space, then p can be used wherever we would use a point in a plane. There are a lot of ways to represent points in a plane. For example, we could put a dot of ink on a piece of paper. Similarly, we can do this.

We can also, thanks to René Descartes, make a data structure representing p as a pair of variables x and y. We could also do polar coordinates r, θ, or any of a variety of other formulations.

Which to use depends on what you are doing with them. The critical element here is you as the person who will be working with and upon the data. Which structuring of variables works for your mind and ways of thinking?

The data structures in your software form the basis of how you think about what you're doing. Data structuring builds and furnishes a home for the res of your work. The more easily your thinking moves through and can settle into the space, the easier it will be for you to work on it.

If the mental building and furnishings are hard for you to navigate and uncomfortable for you to sit and work, you should replace or remake them. If you can analogize a data structure to a room or chair, table, bed, or other bit of household goods, you should be able to develop a feeling for how well it fits you.

If the metaphor doesn't work for you, then you might be able to construct or find another way of experiencing the layout and interactions of your data structures to measure how well they work for you.

There's a thing about particular artworks that don't work for a person. They can be used as inspiration for artworks that do. This very much applies to analogies. You may find the house and furniture image utterly useless, but the manner in which it is useless to you can point toward analogies that do work for you.

DOI: 10.1201/9781003459866-12

There are two kinds of data structures.

The first are those that represent data, things that are purely internal to the program. The sets they range over are internal objects used for tests and calculations and processing of the persistence within the program.

In the furniture and architecture metaphor, these are rooms and corridors and objects that are about the internal shapes and motions of a house. They help in getting around and sitting around. They're chairs, couches, and beds in relatively isolated rooms and passageways that lead to and from the more interesting spaces.

The second type of data structure represents information. They are variables that range over sets outside of the program. These data structures are important in input and output. They let things in and out. They import and export the materials that will be worked upon and played with. In the metaphor, they are food storage, cooking devices, telecommunications, closet space, doors, and windows.

Climate control devices can fall into either category depending on how you see them. If you look at heating and cooling as only affecting the inside, they represent data. If you see them as regulating the differences between inside and outside climate, then they represent information.

Data Structures That Represent Data

These data structures can hold an arrangement of data that fully represents the internal object in all its relevant aspects for all the uses to which you will put it.

The components of these kinds of data structures are all internal information.

By way of example, let's start by considering a structure that represents a point in three-dimensional Euclidean space. Let's further assume that we will need to make numerical calculations from such a point and possibly create such points using calculated values.

This narrows our possible representations since we can't just stick a point on the screen somewhere or mark a dot on a physical object, draw an arrow pointing toward that point, and take a picture of that. We need numbers.

It takes three real numbers to represent such a point. That's what three dimensions mean. Depending on the typing of the language and the accuracy you need, you should quickly be able to figure out the type you need. Most of the time points in space are represented using floating point or double precision variables, but they can be done with integer values as well.

One common method is to structure the point as an array of this kind of number.

This can be quite flexible because the array structure is very general. It's just a sequence of values. Structuring a point as an array doesn't tie you down to a specific coordinate system. But this flexibility means the data structure has no memory in it for which kind of coordinates it uses.

If the array is, for example, [300.0, 3.1, 1.44444472]. That's just a batch of numbers in order. The array does not tell you what the numbers in it mean. It could be rectangular x, y, and z values, spherical r, θ, and φ values, cylindrical r, θ, z values, or any other 3D coordinate system. It could even be providing parameters to a function that will create a point in one of these coordinate systems.

The lack of memory is only a problem if you are likely to get confused as to what you are doing. If you don't need the aides memoire, you might benefit from the generality of this system.

Furthermore, just about every language these days has array functions that can work quickly on this data structure. There's no need to write as many of your own functions to do the basic operations you might need to do with these points.

On the other hand, if you work better with the memory, then you might consider creating a data structure with x, y, and z components or r, θ, and φ or r, θ, and z components so that your points can have rectangular, spherical, or cylindrical representations. You might even create a structure with all nine of these components with functions to calculate the other two triplets from any one of them.

It's likely that one or more of these sound good to you and to the others are either indifferent or annoying. That's the personal taste angle. In terms of computation and storage, these can all be interchanged with only a few bits of overhead (even the nine-component version isn't that much of a nuisance).

Having made our points, let's build from them to create a multidimensional data structure we talked about in earlier chapters to represent an object in motion. We'll call it FullValues.

FullValues needs a 3D value for position, another for velocity, a single real number for mass, and another single real number for charge.

Position and velocity can both be represented as 3D points, so we can reuse whatever data structures we just made. We could do this as an array with two entries or two components labeled position and velocity or p and v, whatever variable names work for you. Mass and charge could also be stored in a single array or given their own component names.

Here is where the grammar of the computer language you are using comes into play. Each language supplies ways to access the components within data structures. The way the language accesses this may feel good or bad to you. Metaphorically, the structure is a chair, and the grammar is the material the chair is covered in. Some access methods may feel itchy or too slick.

One common access method is to separate the structure and its components with a punctuation mark. So, FullValues.Velocity or FullValues→Velocity, for example, are the two different syntaxes in different languages to access the velocity component.

There are also languages where the structure is treated as an array with the names of the components as indices. FullValues["Velocity"] might be the syntax in such a language.

There are reading and coding differences that affect the flow of the programmer. The use of "." as a separator is quick to code but can make the reading a little blurry. "→" is slower to code and can get annoying if you are accessing a component of a component of a component. The "[]" method takes even more keystrokes but if the programmer has an easier time thinking in arrays, then they might benefit from it. These syntactic differences can also be incorporated into the structuring.

If the language uses an array format and you like that, then you might make the position and velocity components an array of points, each of which is an array of numbers. In that context, the y component of the velocity could be.

FullValues["Points"][1][1]

Whereas if one were using the "." component and had each variable separately named, it could be

FullValues.Velocity.y

All of these are matters of personal taste and workflow. Build the furniture that's comfortable for you.

Data Structures That Represent Information

Data structures can represent a selection of data derived from the too much information about an object that exists outside of the system. The goal in making these structures is to have all the data this program will need about those objects.

Don't overburden the structures with unnecessary data, but don't omit necessary data either. Be flexible in creating these structures. As you code, you may find that you need to change what data are represented in them and how they are represented.

Data derived from information tends to enter systems from outside measurements. It may show up as raw measurements or as already-calculated values. Outside sources tend to provide large amounts of data and expect you to pare down what comes in and then place it within your own data structures.

In terms of the data structures, there are four distinct kinds of outside data that one might use in a structure.

1. Access information: This is the data necessary to gain access to the source of the needed data. A URL is access information for remote information. If your system is connected to, say, local cameras or microphones, access information would be how to gather the video or audio feed, generally by providing a pointer to a data stream.

   Access information is usually either labeled by name (such as NOAA for weather reports or Front Door Camera for a local security camera) or by highly technical alphanumeric strings that are readable by someone who knows the particular ins and outs of their local system and are meaningless to everyone else.

   Access information can be combined with the data derived from it in the same data structure, or kept in a data structure used just for the access, or embedded directly in functions that are called to do the accessing.

2. Raw data or data stream. The data that comes directly from the act of accessing can be stored in its own structure or combined with any of the other pieces of information. A data stream that provides a continuous live feed is also a source of raw data. Note that access information would provide a pointer to the stream, whereas the data stream is the data pouring out from actual access.

3. Parsed data. This is the data transformed from raw/streamed to the form that the sender thinks you need. Parsed data is going to be somebody else's idea of what data you should be working from. Continuing the furniture metaphor, this is like a piece of furniture or an appliance that came with your home and you can't replace, that has its own methods that can be annoying to you, and that you have to keep

using it because no one makes things the way you need them to, and why can't you just make your own National Weather Service?

Parsed data will often lack what you directly need or will have it stored in a data structure that was built for someone else's ways of thinking or multiple people's ways of thinking, none of which quite work with yours.

4. Calculated data: This is the data you have calculated from the raw or parsed data in order to create the data your program actually needs. The same principles that apply to creating data structures for data apply to making calculated data structures. You will likely need to bend and twist a few of your ways of thinking to more easily connect the externally parsed data to the interior needs of calculated data.

If your program is creating data that will be supplied to other programs accessing your software, this sequence of structures, access→stream→parsed→calculated, will need to be reversed.

You will need to begin with the data you are calculating. From that, you will need to create a parsed form suitable for travel. You will then need to pack it for travel using whatever methods of data transport you are set up for, and you will then need to provide means of access for others.

The pivotal problem in making this export and import of data work occurs between calculated and parsed data.

The calculated data are idiosyncratic to your own system, your forms of data generation, and what works for you personally.

The parsed data are for the usage of others, a diverse space of people, each with their own needs.

How can you determine which data to generate and how to pack the data in a way that does not require that someone think the way you think?

The most common solution is to use whatever form of data packaging is fashionable. There have been a number of such methods. XML and JSON have had their times of being in vogue. SQL databases are in quite common use at this time. Each of these methods will use your own organizational logic encoded in someone else's idea of the logic of organizational logic.

You can also write your own parser and let others use it. This anonymizes your personal storage and calculation processes but still presents things as you think they should be presented.

You can use a multipackage system that allows people to download different bundles of information as they need, although this presents the problem that people will have to work through how you organize what you organize.

The most effective method is to illuminate why you organize things as you do and how you expect your data to be used. You can also lay out tools and parsers to get other common forms from that and perhaps provide customizable tools that will allow your accessors to more easily go from your idiosyncrasies to theirs.

If this sounds like the basic problems of artist and audience compatibility, it's because that's exactly what this is. Data structures are embodiments of how programmers are comfortable thinking and working. My home may not be comfortable for you, nor yours to me. I can try to make my arrangements more convivial for you and vice versa, but that doesn't always work. Social discomfort happens.

Having to once or twice be in an uncomfortable awkward place having a difficult conversation with someone whose thinking you just don't get is unpleasant but livable.

If it looks like you would have to do this again and again, then you should see if there are alternate sources for the data you need or alternate ways to deliver the data you have. Long-term iterated frustration can wear down everyone involved until eventually people are just sticking their heads out the window and guessing what the weather is going to be rather than spending another minute digging through predictions of a chaotic system.

Structure and Infrastructure

Some data structures are built to contain the data representing the outside visible directly measured aspects of things, the dimensions of walls, the colors of objects, the materials of upholstery, and so on. We will call those structural data structures in contrast to infrastructural data structures that have the joints and nails that hold things together, the plumbing and wiring within the walls, etc.

The structural data structures have data that directly affect the user's experience and awareness. For example, in a roleplaying game, a character's character sheet is a displayed data structure informing the player of what their character can do. The user will repeatedly directly examine the contents of the data structure through the interface and determine what actions to take in a given situation accordingly.

The data used to draw the characters and animate their actions and interactions with the virtual environment are infrastructural. They are not directly presented but supply the underlying data for what is happening within the program.

The data you use to draw with is infrastructural. The drawings produced are structural.

Structural data structures have a direct audience presence and need to be treated accordingly. They cannot be purely idiosyncratic. Infrastructural data structures just need to work for the things you are doing with them.

There are two basic methodologies about the relationship between structural and infrastructural data: Supplied need and shared components.

We'll use plumbing as a metaphor/example for both of these relationships.

In supplied need, the infrastructure needs data from the structure and/or the infrastructure from the structure.

The structural and infrastructural elements in the supplied need are coded as separate data structures without any components in common. What one needs is calculated from the other and taken inside it.

If the structure is a faucet and the infrastructure is behind-the-walls plumbing, we would like to know how much water will flow out if we turn on the faucet. When the faucet is turned on, the water flowing through the faucet is calculated from the values in the plumbing infrastructure, such as water pressure, and assigned a value in the faucet data structure. The datum of water pressure is only present in the plumbing. The datum of water flow is only present in the faucet.

In shared components, the structural and infrastructural data structures have components in common. Changing the values of one of those components through either the structure or the infrastructure affects both of them.

The water flow value could be a component both of the faucet and the plumbing. Changing the water pressure in the plumbing would change the water flow component in both, as would turning the faucet on and off.

Any programming problem with structure and infrastructure can be solved with either of these methods. Which to use is purely a matter of what works best in your thinking, whether you have an easier time seeing connections as steps from one thing to another or as aspects shared between them.

Recurrence, Inheritance, and Uniqueness

Data structures are created because a given program will frequently find itself needing the same kinds of data over and over again.

When a data need shows up only once, it's usually more efficient to calculate it and use it immediately.

When there's more than one thing that has that arrangement of data, it's often better to create the abstract organization of the necessary kinds of data and then make individual instantiations of that structure by filling in the necessary data elements.

This form of shaped recurrence is vital for most human concepts of what objects are. We have concepts that can be partially perceived and conceived by filling in the entries in a set of values.

Note that this is not as true of information as it is true of data. A chair can be specified by giving dimensions and materials and an idea of the shape as a description but a real chair has much more entangled information than the data to describe it.

A one-and-a-half-meter-tall wooden chair with a straight back and casters covers a lot of possible chairs. It's enough to classify to a certain granularity, but not to actually encompass the whole real object.

Human categorization is a way to classify what one is talking about but does not fully present the real thing. We often try to solve this problem by adding subcategories or adjectives. There are chairs. There are office chairs. There are office chairs with cup holders. There are green office chairs with cup holders.

Data structures can increase in elaboration through a process of inheritance which is the equivalent of adding adjectives.

If you have a data structure for a chair, it's not necessarily advisable to fill that structure with all possible characteristics of all possible chairs that anyone has ever made, for two reasons.

First, that will be too data-heavy and have too many characteristics that do not apply to all chairs. A chair with wheels has a bunch of data about the characteristics of those wheels. Having the wheel components listed for a chair without wheels unnecessarily bulks up the chair data structure.

And, second, you may think that you've listed all the components of all possible chairs but tomorrow someone might invent a whole new kind of chair with its own added characteristics.

Inheritance solves this problem by having a wheeled chair inherit all the components of the chair, but adds the components of wheels. An upholstered chair can also inherit from the chair and bring its own variable. Some languages allow for multiple inheritance so you could have a wheeled upholstered chair that is both a wheeled and an upholstered chair. Other languages require single paths; so, wheeled upholstered chairs would have to inherit from either wheeled chairs or upholstered chairs.

With enough inheritance, it becomes possible to have data structures to whatever granularity you are using for the recurrent objects you are using.

There is an interesting phenomenon about recurrence with enough specificity of data which is that enough recurrence with variation creates uniqueness as a consequence without actually specifying that things are unique.

Even if one is dealing only with data, not information, it is possible, easy in fact, to create unique values for a single instance of a data structure.

It's a matter of probability and distribution.

If you have 20 values in a data structure, each of which can take one of ten values, you've got 10^{20} possible configurations. Assume that you will only ever need a hundred, a thousand, or a million of this data structure. If you randomize the values with an even distribution, the odds are that a million results will have no repetitions within it.

You have generated uniqueness by having a sparse number of results relative to the number of possible results. If, on the other hand, you needed 10^{100} results, then you would be vastly overwhelmed with duplicates.

This may sound simplistic, but consider the following. Human reproduction involves the creation of haploid cells, which, barring non-disjunction, will each have one chromosome out of each of the 23 chromosome pairs of each parent. That's 2^{23} possible chromosome sets for each parent, which is 2^{46} possible resulting pairings. That's around 64 quadrillion possible genetic combinations from just two people. Except in cases of identical twins, triplets etc., the genetics of each human being are unique.

And that's just the genetic data. It doesn't even get into the complexities of information in fetal development, growth, life experience, and so on. Human genetics creates uniqueness just by using probabilities. Human life expands on that uniqueness, creating broader and broader divergence of possibility.

Coevolution: Architecture, Furniture, and Clothing

There is an interesting phenomenon in art that furniture and clothing develop together. What we sit on and sit at has to work with what we are wearing when we sit down.

Furniture mutates from large-scale changes in clothing fashion and fashion can be pressured by furniture. People have developed many different ways of sitting when their clothes and furniture don't work together. Those people will also look for new kinds of seats when they go shopping for furniture.

Furniture is under pressure from the buildings it needs to go in. We need to be able to move through the houses and get to the things in them. People need to be able to sit and move and work and play and store and retrieve from the furniture. Their clothes and their rooms both push at the possibilities of furnishing.

If we continue the metaphor of data structures as furniture, we see the pressure of how you the programmer wish to arrange the data structures in the architecture of the program. There is also the pressure of the actions that will be moving the furniture and moving on them, in them, and around them. These actions are the clothing of the code and the users.

Leaving the metaphor behind and coming back to the facts of programming, we need our data structures to fit the functions that will use them as parameters, modify them in operation, and produce them as results.

Personal Exercises

1. Skim through any math, physics, or programming text you've used in the past. Spot which particular ways of thinking about particular variables or sets of variables have been easy for you to work with and which bogged you down or confused you.

2. Look at the data structuring style of at least three different languages. Consider which forms you like and which you dislike.

3. One of the impossible tasks of data structuring is representing uniqueness, especially human uniqueness. It is not possible to measure a human in all aspects, nor to generate a computer representation thereof. But it is very often necessary to create an internal stand-in or agent for a person. Do testable exercise 3 and then consider what is lacking in the data structure created.

Testable Exercises

1. Pseudocode a data structure with at least six components. Then code the structure in the three different languages you looked at in personal exercise 2. Make note of your preferences and why you like what you like.

2. Take an online information source you use but do not like. Describe the data it provides and what you need to do to make use of it. Pseudocode a data structure that would provide the information you would like to receive.

3. Imagine a program you might like to write that would need to hold and access information about human beings. Pseudocode is a data structure that has that needed information. Explain why the program would need the components of the structure and why it would not need more information than that.

11

Functions and Objects

Functions and Parameters

The mathematical definition of a function is surprisingly useless for most programming purposes. A function is a set F of ordered pairs {(a,b)} such that if (a,b) and (a,c) are members of F then b = c. It's essentially a minimalist definition that lets one write F(a) = b as a different way of saying (a,b) is a member of F.

From a programming perspective, a function defined this way is a possibly infinite lookup table. That's not what we mean by function. We mean a process that we can code to take parameter values and generate results. Let's dig into that.

When functions first crop up in math classes, they are presented as calculation instructions like $f(x) = 5x^2 + 3x + 2$ or in word form like "the sine of an angle in a right triangle is equal to the length of the side opposite the angle divided by the length of the hypotenuse."

In both of these cases, the function is presented as operating on a particular value (x in the case of the f(x) above and the angle in the case of the sine) called the parameter of the function. Then the function is presented as work that needs to be done either by hand or by computer to produce a result. It is also presented in a finite form. The infinite lookup tables are replaced by performable processes that can be done in finite time and presented in finite space.

The work is what we code and the work is where most of the art in making functions is found.

Before we dig into that work, let's look at the parameter (or parameters).

What types of things can a particular function operate on?

Parameters are a particular use of variables, and as we said before a variable can take any value in a specified set.

If we look at a function f where f(a) = b, a and b are both variables, the set from which a takes its values is called the domain of the function and the set from which b takes its values is called the range.

In programming we simulate domains and ranges with type specifications. That is, we say what types the parameters can be and what type the results can be.

From the pure math perspective, all functions between a given domain and a given range preexist because you can generate the set of ordered pairs of the domain and the range and the functions are just the subsets of that set of ordered pairs where each domain element has only one corresponding range element.

DOI: 10.1201/9781003459866-13

But again, that's not the work of calculation. From a programming perspective a function is coded. It's an artistic creation that takes parameter values, operates on them, and produces result values.

So, while the theoretical but empty of process functions pre-exist, the created artworks that are individual processes of embodying a function need to be made in order for them to exist and be usable by people.

Every function you code that takes certain types as its parameters increases the set of things that can be done with those parameters. Crafting functions increases the functionality of the parameter sets.

From an art perspective, the more techniques there are to work with something, the more material-like that something is.

And this is true in mathematics. The objects of study in abstract algebra are defined mostly by the operations adjoined to sets. The integers are just a countably infinite set if they have no operations. Giving them addition makes them a group. Giving them addition and multiplication makes them a ring. Giving them < makes them a well-ordered set and so on. The integers are easy to work with because so many ways of working with them have been crafted and used.

Each function you make increases the space of interactivity between parameters and opens up new possibilities for results. That interactivity can be seen as a creator of entanglement of information within the data.

$1 + 2 = 3$ is an entanglement between 1, 2, and 3. $2 * 3 = 6$ is an entanglement between 2, 3, and 6. $2 * (1 + 2) = 6$ follows from these two entanglements.

If we add the general entanglement $a * (b + c) = a * b + a * c$, we get $2 * 1 + 2 * 2 = 6$, which gets us $2 + 4 = 6$.

The functions create interrelations which give us things we can reason about.

Our sense of what things are is largely dependent on what can be done with those things and what we can deduce or imagine doing with them.

This points toward creating functions that make sense for what their parameters are supposed to do. The datatype string was created to represent what we read and write. A string is often implemented as an array or linked list of characters (usually ASCII or Unicode characters). The flow of language coming from the back of mind through my fingers tapping on the keyboard is being turned into strings by the computer I'm writing this on.

Assuming we can read and write them, what else do we do with strings to make them more like text?

Ask any writer or editor and they'll tell you eventually that text needs to be editable. You need to be able to change it coherently. Ahem, anyway.

Editing operations have been coded that take strings as parameters, like the ability to concatenate strings, delete parts of strings, replace part of a string with something else, search through a string to find particular substrings, and so on. The functions make the strings feel more like what we're trying to have them model, which is to say editable text.

The being able to read and write them is a lot more work to code because human reading and writing is a connection between the data of the strings and the information within human minds. We can't get computers to do that, but we can create presentations of the strings that make it easier for humans to do the actual reading.

Functions have been made that take strings and fonts and from them produce graphics that show up on screens as typeface and other functions that produce sounds from strings that sound like someone reading the strings. We also these days have functions that can transform strings into sounds that our ears can interpret as language if we know the language involved. These functions strongly connect the infrastructure of strings with the structural presence of text. Once those are connected, we have computer data that entangles easily with our human thinking about text and meaning.

Functions of meaningful parameters that produce meaningful results need meaningful calculations. The process needs to make sense with what it's working on to produce what comes out of it.

There's a feedback loop here. If you make a process to go from something meaningful to something else meaningful you are also generating a sense in a person following the process that the process itself makes sense.

This involves some risk, because people don't have a direct sense of what makes sense. We have a combination of fundamental ways of thinking, cultural and personal teachings and life experience that gives each of us a distinct concept of what making sense means. Some of these correlate better to reality and human life than others do. It's why we also benefit from developing the ability to examine what we think making sense means and check it against how well it works for us and for other people.

Most human-calculable functions use relatively simple sets for domains and ranges.

In computers we can have quite complex data structures with lots of data in them for parameters and results.

We need to make our data structures and our functions work smoothly together.

Let's go back to our example of various ways of implementing points in three-dimensional space.

Suppose we need to produce a function for the distance between two points using the standard distance function.

That function arose from the Pythagorean theorem and is optimized for rectangular coordinates. In that form, the process can be described as follows:

1. Calculate the differences between each pair of corresponding rectangular coordinates.
2. Square each of the differences.
3. Add the squares.
4. Take the square root of the sum.

If your implementation of the data structure point uses an array of coordinates you can loop through the array, calculating and squaring and adding to a sum, which you then take the square root of.

If you have the points as x, y, z values, then you'll likely need to calculate each of the coordinate differences separately (unless the language you're using lets you loop through components by name), then add them up.

If you are using spherical or cylindrical coordinates, you'll likely need to calculate x, y, and possibly z values from the length and angle values, then do the calculations above.

You might do this by creating another function to calculate rectangular from polar and call that function once for each point within the distance function. If you end up doing this a lot, it can be more convenient to have your data structures have both rectangular and the other coordinate system values within it.

It's often more optimal to store redundant data than do a lot of calculation.

A lot of work has been devoted to optimizing the calculations in functions. These considerations should be applied especially for functions that are called frequently in your programs.

Some of this optimization is done in compilers and assemblers. You needn't spend your time trying to eke every last cycle and squeeze every last byte down the minimum, as once was necessary.

It's good to learn how to optimize functions. It's wise to learn when optimization is needed and when it isn't. Don't break your own ability to read and play with your code in order to fit optimization.

Optimization is not always the optimal way to make a function work. There are insights that can arise out of non-optimal calculations if you play with them.

There are whole books full of problems in calculus that involve using non-optimal formulations.

These problems commonly take the following form:

Find the integral of f(x)dx for some function f which has no easy integral.

And yet f(x) = f(x) + 0 and 0 can be anything minus itself. So if you can find a g(x) such that f(x) + g(x) is easy to integrate and g(x) is easy to integrate, then the integral of f(x)dx = integral of f(x) + g(x) dx - the integral of g(x)dx.

In effect the problem is solved by adding zero in the complicated form of g(x) − g(x).

Figuring out what g(x) needs to be is part of the art of integral calculus.

And yes, that's an art.

We're relying here on a = a + 0 for any a which equals a + b − b for any b.

We can also pull a similar trick with multiplication.

a is equal to a * 1 for any a which equals a * b/b for any b ≠ 0.

That last anti-optimization shtick is used in complex numbers to define complex division.

Suppose we need to calculate (a + bi)/(c + di) .

That will equal ((a + bi)/(c + di)) * 1.

If we express 1 as (c − di)/(c − di), we turn the problem into:

```
((a + bi)/(c + di)) * ((c - di)/(c - di))
```

which equals:

```
(a + bi) * (c - di)/(c + di) * (c - di) .
```

We picked (c − di)/(c − di) because (c + di) * (c − di) is the real number $c^2 + d^2$ and we can already divide a complex number by a real number (just divide each of the real and imaginary parts by the real number).

So we end up with $(ac + bd)/(c^2 + d^2) + i(bc − ad)/(c^2 + d^2)$. Messy but meaningful.

In a sense these methods involve doing nothing in a complicated but useful fashion. After all, adding zero and multiplying by one do nothing. They do not transform the values they operate on. But they can reveal those values in new lights from new perspectives wherein the values make more sense.

Be aware there are an arbitrary number of problems that can be solved by doing nothing in a complicated fashion.

A great deal of human problem-solving and art come from seeing the possibilities of what can be done within the space of doing nothing.

Functions as Invocable Code

Let's shift perspective from functions as math embodied in code back to the programming idea of functions as invocable chunks of code.

Invocation can make code more readable and more meaningful, but it can also make debugging more convoluted.

Code with invocation resembles how humans converse. Much of what we say are references to ideas and methods shared. This is especially true of the pseudocode that people use to give instructions and lessons.

Recipes are invocations of what the cook reading the recipes is supposed to already be able to do (separate four eggs, whip the whites with ¼ cup confectioner's sugar until stiff, fold the whipped egg whites into chilled chocolate ganache).

Each of these instructions can be written in a function and parameter form.

Separate(four eggs) = four egg yolks, four egg whites

Whip_until_stiff(egg whites, sugar) = meringue

Fold(a,b) = a wide diversity of interesting results depending on materials.

If the person reading the recipe does not know how to do any of the processes, they can look up and learn the methods then implement them. The invocation of a not-yet-implemented process requires that the process be implemented, either by a human learning or code being written for a computer or both.

This metaprocess of breaking down a needed process into invocable subprocesses is an elaboration of the calculation metaprocess discussed in Section I, Chapter 5.

Now we can fold in the understanding we've developed since then. We have a better understanding of the data and information that make up initial parameters. We also have a better understanding of what data and information we need to emerge from the final step.

We can examine the beginning and end phases with greater clarity because we can use the framework of data, information, and data structures. We can more clearly organize what we're working from, working with, and working toward.

Making our procedures from the middle between parameters and needs, we can build in three aspects of our functions: The main body of the process, the side effects, and the returned result (if any).

Main Body

There are a number of potentially useful metaphors for the main body of a function. These metaphors can make it easier to see the code being called upon as both separate from and interacting with the part of the program that invokes the function.

Standard programming texts implicitly employ such a metaphor when they use the term "localization." This phrase sees the main body of the function as a place, a locale where things happen in isolation. In this metaphor, the parameters are imported to the main body and the results and side effects are exported from it.

The locale metaphor is a good one for many people. It gives the sense of the function working in separation from the code that invoked it. But locales can still access things that are available everywhere, such as the input and output devices, events, file storage, and common libraries of functions and data structures.

Many classical languages also allowed the ability to declare global variables which were always accessible. This fell out of favor because global variables are too easy to change in multiple places. Bugs that developed because one global variable was changed in some obscure corner of the code are notoriously difficult to track down.

Another useful metaphor is our explication of functions as means of entanglement. This sees the functions not as isolated places but as infrastructure within the interaction of data structures. This metaphor will be extended and elaborated when we dig into objects later in this chapter.

A third metaphor that we deliberately put into how we talked about branching through functions is that of invocation. A function is called in like a specialist to handle a specific kind of problem. In this metaphor, the function has a certain style and method of action and is called because the style works for the job being done.

A fourth metaphor is function as tool. In this a function is seen as an object to be picked up and employed by something that uses it rather than as a character called in to do a job.

It is quite possible to combine the previous two metaphors, with some functions being seen as characters and others as tools those specialists use when called in, a doctor and a doctor's bag, a carpenter and a hammer, and so on.

A fifth metaphor goes back to the awareness of invocation as branching and a program as a flow of processes. In this form, the function called in a particular place is like a branch of a river that may take one briefly away from the main flow and then return to it later.

Whatever metaphor (whether one of the above or another we didn't think of) that fits your style of work and coding will inform how you construct your functions and what awareness will be necessary in employing the function.

The way you think about the functions will frame how you create the processes of the function as code separated from the main body of the program. It will also affect which bundles of functions you create to handle certain tasks that the main body and other functions will call upon.

The metaphor also affects how you see data structures in relation to the function as parameters, recipients of side effects and as return values.

You do not need to have a single metaphor for all functions. I use all of these ways of thinking about functions depending on what role the function plays in the code. Some I see as places, some as entangling operations, some as called-in experts, some as tools to be picked up and used, and some as parts or branches of a river. And sometimes I see the same piece of code according to more than one metaphor.

One of the convenient aspects of this multi-metaphor approach for my personal style of coding is that I can treat each part of the program as how it seems to me. Its res will have a shape and flow that will fit one or more of these analogies. I can work from that to the dictum of the code and back from the dictum to that analogy to see if the code is doing its job in the way it needs to be done.

Having a sense of how the main body of a function makes the way that the function needs to operate can ease the process of building the code and make it more enjoyable.

There is an exhaustion that comes from simply sticking one bit of standardized optimization after another to produce what should, in theory, work, but which you cannot make work for you.

Getting a feel for the functions as you are planning and making them can make the work more effective, more enjoyable, and more capable of fitting into the rest of the program.

Side Effects

Side effects are the lasting consequences of the function to the persistence of the program that are not embodied in the returned values.

There is an attitude toward side effects that flows in and out of fashion, that functions should not have side effects at all. Those who follow this fashion assert that the only transformations and consequences for a function should show up in the returned value or values.

Object-oriented programming will often make an exception by saying that the object that an invoked method belongs to can be changed by the method, but other objects should not.

Both of these attitudes emerged from the no global variables aesthetic.

If we are not embracing this aesthetic, we can ask what side effects this function should have.

If we go back to our five processes, we can see that side effects can be storage that is not local to the function, possible branch invocations of other functions, input required or allowed, or output produced in the course of the function.

Side effects can therefore be storage of some data outside the local scope of the function, temporarily leaving the function for another function that is not directly involved in the actions of this function, gathering data from users, or presenting data to users.

These can be combined. For example, a user might need to supply an action that will lead to an invocation of another function which might lead to a display of effects to the user and possible storage of consequences of that invoked action elsewhere. This is very common in games where user actions and game reactions are the fundamental loop of the program.

It can help to think of side effects as the function reaching beyond its scope to gather more in or place more out.

It's a good idea when doing this to be aware of/make note of what a function is affecting beyond itself, so you can see the flow of consequences and changes.

If it works better for you to minimize side effects then do so, but do not feel constricted by the idea that you must do this. Code as works for you. If your functions are bouncing around and reshaping the environment of your app and the app works better for doing it this way, and also if you enjoy building code that way, then do it.

Return Values

The last consideration of a function is its return value, if any. Some languages distinguish a function (invocation with a return value) and procedure (invocation without a return value). We'll just stick with function as the term for both.

Return values are useful if the main body and the side effects don't do all the work. In particular, if the function needs to produce something that will be employed in some other code in the body that invoked the function, then there is a benefit to presenting that needed value to the main body.

Obviously, mathematical functions and many string manipulations need returns. Functions that construct data structures and objects need to return what they constructed, as do functions that pull data from files or off the internet.

Functions can operate as tests and return Boolean true false values.

Some of these test results are used to confirm or deny whether the function performed some side effect. For example, a function to save data to a file might return true or false to say whether it successfully saved.

Most tests are calculation tests. There's an aesthetic variation between preferring to have the test embedded in the function or to have the function return a value to be tested against. If checking whether the result of a calculation is equal to a particular number, do you have the calculation and the test be within the function so the return value is true or false, or do you have the function do the calculation and then check for whether the results are equal in the code that invoked the function? Either approach works.

One function can perform multiple tests on its parameters and supply the results as an array or other data structure.

More generally, multipurpose/multiresult functions are handy if there are a number of different pieces of data that can be derived by doing a number of operations within a single loop.

There are quite a number of statistical calculations and/or data examination operations that can all be done by iterating through a set of database records or other forms of raw data storage.

Large long loops for multiple purposes are often better done as one multipurpose function that returns a data structure of amassed calculated information rather than using separate calls.

For example, a single function to provide maximum, minimum, sum, mean, median, standard deviation, and perhaps geometric and harmonic averages of a large dataset might be of more use in an analytic program than having those as separate individual value functions, each of which would need to loop through the dataset to determine one number.

Let's go back to the relationship between function and parameters and turn the matter around to see the functions as aspects of their parameters instead of operators on them.

It's time to look at object-oriented programming.

Data Structure + Methods = Objects

From a language design perspective, object-oriented programming can be seen as a way to organize the coding of data structures and functions so that code for the functions that work on each data structure is written right after the declaration of the data structure. Object-oriented coding makes it easy to refer back to the structure, see what you have to work with in each function and keep the whole thing together in your head.

But we're talking about human heads here, where what you put together becomes associated. A freestanding function invokable from somewhere does not tend to create the same mental structures as a method explicitly coded for an object type.

From a mathematical perspective, the object-method entanglement of data structure and functions both makes sense and is a little off.

It makes sense to associate integer addition with the integers, real number addition with the real numbers, vector addition with vectors, etc.

But we tend to look at addition as a binary operation, a + b, rather than a method Add(b) that is given as an instruction to a. a + b fits the aesthetics of algebra better than a->Add(b).

This can get mentally twitchier when you consider something like rotating a vector using a rotation matrix. If you're coding this process, you might have a vector object and a rotation object (or a general matrix object). Do you make RotateBy(matrix) a method of vector or Rotate(vector) a method of matrix?

And here's where the art sneaks in. While these two formulations are mathematically equivalent, they are artistically different because they focus the mind on different aspects of the situation. Are you following the flowing changes of the vectors as they get rotated or are you watching the actions of the matrix as it transforms the directions of vectors?

From a writing perspective, the question is which character's point of view are you looking from? Do you follow the hero (vector) turned by the twists of fate (matrix) or do you look from the divine perspective of the matrix changing the fates of many self-defined heroes?

The object-method paradigm brings programming closer to fiction writing because it gives active character to otherwise inert data structures.

What can this object do with these other objects and how does it do it?

This is not so easy to see if we look at an Add method (although integer addition, real number addition and vector addition each have their own character), but it is clearer if we have an object representing a CGI model and a method Dance (this_dance).

Each model can have its own ways of dancing various dances coded into it. The particulars of that Dance method for each kind of model will illuminate the character that the model represents.

While it may seem a leap to go from addition to dancing, this is the utility of anthropomorphization, an aspect of human thinking that shows up in all arts at all scales.

Anthropomorphization is usually described as perceiving and acting as if something not human is human. I find it more useful to think of it as seeing non-human beings and things relative to humanity and the ways humans do things.

Saying that clouds dance or raindrops drum or volcanoes rage gives us a space to work from to get a sense of how these things act. If we have a cloud object and see part of the way the cloud moves and changes shape as dancing, we could give a Dance method to the clouds.

We might copy a Dance method that we were using for a human figure into the cloud object, which would look weird and humorous, very much like a classic style of animation in which non-human things are absurdly humanized. Or we might consider the mathematics of clouds in winds and create a Dance method that has the chaos of actual weather evolution as the underlying motion of dancing clouds.

Seeing the cloud in action as if it is related directly to human action can lead away from reality to humor, but it can also lead directly into a model of reality. The humanization caused by seeing a cloud as dancing can give a human mind more ability to play with the underlying reality.

We've seen this before where a highly humanized activity like gambling can help explore a seemingly inhuman understanding like quantum mechanics.

Crafting methods for objects is a process of exploring the character of the object because of what the object will need to do and be used for. This is like fiction writing. Character is revealed in action, interaction, and introspection.

Clearly, I made this connection because I code and write and think the way I think. So am I seeing a connection that's there or projecting a connection that works for me, or some of both?

Some of both, clearly. The fact is all art involves humanization because all art made by humans for humans has to relate to humanity. And character as a concept is not confined to writing. Every artwork can be seen as having a personal character to it. Developing the character of one's work is part of the sense of making its dictum and res come together.

Inheritance and Taxonomy

When we discussed data structures, we spoke about inheritance, wherein a structure can be another structure with more stuff added. Objects inherit both structures and methods and allow more components and methods to be added. They also allow methods of a class to be replaced in that class' subclasses if the child object does things its own way.

If the parent object has a Dance method, a child object can either keep that method or create its own way of dancing that works for it which might or might not invoke the parent method. The instruction to dance will invoke the method for the object itself. If it doesn't find one it will go one step up the inheritance ladder see if the method is there and invoke that if it is. If not it goes up further.

Languages with multiple inheritance can get tricky if two ancestors on distinct lines each have the same named method. Check the grammar of the language you're using because they don't all resolve this kind of problem in the same way.

Object inheritance is sometimes used to replicate biological taxonomy. One might have a broad abstract object type like living_being and give it subtypes for each of the domains, then kingdoms as subtypes of the domains, phyla as subtypes of the kingdoms, then class, order, family, genus, and species.

This kind of taxonomy can be overdone in object-oriented programming. The need for it in biology is one thing, but while the higher-order divisions have distinct qualities it's hard to see them as having anthropomorphic/human-comprehensible character and methods. You have to go pretty far down the tree of life before methods of living, acting, and interacting really crop up.

It can sometimes be better to have the paths of taxonomy be properties within an object rather than be a tree of inherited properties. One might start the living_being types at class, order, or family and have domain, kingdom, and phylum be properties within the data structure of the object.

I find it's a good principle to start your classes where they have enough interacting qualities and methods to feel like they exist as distinct presences, each with their own persistence and actions. But if the full taxonomy or some other variation works better for you, use it.

Instantiation

Object-oriented programming puts object and actions together and localizes them to each instance of each object so that actions and reactions are seen as "what happens to this thing when it does this" and "what does this thing doing this do to these other things."

These questions bring the human concept and grammar of doing to the more abstract space of functions, parameters, and results.

Most people have an easier time thinking in terms of objects doing than functions doing. But functions also have distinct and playful character and for those who can see and play with those characters, function-oriented programming has many of the advantages of object-oriented programming. People who can see both the character of objects and the character of functions can go back and forth and combine both kinds of programming.

Use the kinds of programming that work for whichever way you see the actions you are coding. Perceiving character in things makes it easier to work and play with them, so try to do the kind of coding that gives you that kind of living awareness of what you are making.

Clothed Motion

Between the extremes of f(a) = b with no illumination of how f takes a and makes b, and a full-on lushly choreographed, costumed, and staged dance number lies the question of what to show of a function and how to show it.

There is an art that is best at showing aspects of motion. That art can serve here as analogy. We're talking about the art of humans wearing clothes. So, yes, we are harkening back to the clothing metaphor for data structures and now we're showing why we used that metaphor in the first place.

Clothes affect what motions are easy or hard for an individual human body to make. They emphasize and deemphasize aspects of those motions.

This transforms the question of how to show the function into the more metaphoric question of how you clothe the function for different occasions.

When debugging, we can reveal the process step-by-step or mark lines in the code that we wish to pause so that we can see what is happening when those steps are reached. This is clothing the function in a highly revealing functional suit of clothes that shows whatever level of the underlying movement we need to determine what exactly is going on and how fast each part is happening.

This is clothing for time and motion studies and safety inspections. It is purely for our own needs. Most programmers hope to never show their work like this to anyone else because it is very slow and clunky and does not show the character of what is going on.

Clothing a function to be illuminating to users depends on what you wish to illuminate.

Let's go back to rotating a vector function; the interior code will usually involve multiplying vector and matrix components, then adding them up and ordering them to create a new vector. The interior motion is the algebra used to efficiently calculate the rotated vector. This can be even more efficient if the vector is expressed in spherical coordinates and the rotation is expressed as adding angles. But again, that's algebra.

Rotation is something that has a direct meaning for humans and that meaning is a geometric one. We rotate ourselves by turning in place. We rotate objects in our hands.

To clothe the motion of the rotation in a way that will relate the algebra to the user's human sense of rotation, we might need to calculate the resulting vector, then generate an output of the effect of rotating through a succession of small angles between where the original vector was and where the final vector is. We could do this by showing a point in motion, an arrow turning like the hand of an analog clock, or by showing an entire virtual world being rotated through by a virtual viewpoint that was looking along the initial vector and turned to look along the final vector.

All of these clothe the motion that the function is calculating though they do not show the actual calculation.

This is a kind of stage magic where we are doing one thing, but we act like we're doing another.

Stage magic is an art, and its illusions can be revelatory of the function space underlying actions taken.

f(a) = b is itself an invocation of a large space of possible ways to get from a to b. Each of those ways is an embodiment of how to make that passage.

Artistically, there is the difference between the composition that the artist uses to put the work together (the coding in our case) and the composition that is shown to the audience. These compositions are entangled but not the same.

Palette to painting composition creates what is there to be seen. It is artist composition. A member of the audience looking at the painting composition with human eyes that recognize figures and imagine their past and future is seeing what this person sees in what can be seen.

Clothe the functions in ways that reveal how they actually work if the user needs to see the actual work, but if they don't need that, then clothe the functions in ways that reveal what the art is trying to show to the audience.

Personal Creation

When building functions and objects for programs you are making yourself, it can be beneficial to delve into how the processes you care about work for you.

Even if no one else sees the code, you need to have what interests you be alive for you.

This does not mean that you should try to make everything matter to you. There are functions you will likely need that are not to your taste. Such functions will usually have efficient but not necessarily interesting methods of coding that other programmers have shared online or put in books. There may be whole APIs with many available chunks of code for related functions.

Adapt these imported methods to your needs and ways. If need be, build functions that take the way you do things, translate that into what the imported function needs and build translators back from their form to yours. This can be a nuisance and you may find that you need to dig through the assumptions others have made in order to find something conformable to how you work things.

Sometimes you can find yourself staring at a standard method that does not fit at all, but the problem of how to make something that does work for your program does not seem interesting to you.

If so, you should have a look around at alternative methods. If none seem to work for you, you'll likely need to build something for yourself.

All of these alternatives can be annoying. It's tempting to just get it over with and go on to other things.

This can work.

But if it doesn't, if every method you can find is annoying, something else might be going on.

Uninteresting-but-irritating can be related either to something that is interesting to you from a different perspective or something on a larger or smaller scale than you are looking at. The better artist in the back of your mind can express themselves through irritation.

Bear in mind that it's rare to be the only person bothered by a particular thing. If you solve a problem that few people talk about as a problem, you may discover that it bothered a lot of people, they just had no way to talk about it as a problem or show anything that could be done about it.

The human mind works upon need and awareness of need.

Sometimes need shows up as beauty one seeks to make or good one seeks to bring about or truth one seeks to discover.

Sometimes it shows up as things that bug the heck out of us and we work to get rid of.

If the only solutions to a problem that you can find don't work for you, maybe the work you need to do is create solutions that do.

Group Creation

What if you are not working alone?

So far we've focused on working alone or working on parts of programs where you are responsible for one aspect of it and others will interact with your work through functions and objects you supply, and you'll interact with theirs through functions and objects they supply.

What if the construction of the data structures, functions, objects, and interfaces are not separated but made within a group of people? What if not just the dictum but aspects of the res are entangled between a batch of interacting minds?

The tumultuous history of musical bands shows that this is not a simple or easy task to deal with. We will delve deeper into the subject in Section III, Chapter 19. Right now we're going to focus on the specific process of how to create functions and objects out of multiple perspectives.

Let's once again invoke the elaborating from the middle metamethod.

What do we have?

How do we get what we need from what we have?

What do we need?

It can quickly emerge that a group of people presented with this triad of questions can disagree about all aspects of the first and third questions, and arguments can break out over beginning and ending long before they get near to working the middle.

This happens because both what we have and what we need are matters of information, of which there is too much.

All people involved bring their own personal TMI to the discussion.

This can work out well if all people involved are aware that they bring their own viewpoints and artistry to the situation.

Some of that TMI can be the awareness that the initial framing of the problem and the initial idea of need are too narrow.

One person may think, for example, that despite having terabytes of gathered data, crucial facets of things or entire regions or groups of people have been ignored in the gathering of that data.

Amount of data is not an adequate measurement of sufficient data.

Similarly, needs can be too narrowly defined by assuming that everyone who might be using a program has particular capabilities or kinds of hardware or internet access or personal attitudes.

It's important, therefore, when considering the initial awareness of the problem to listen to the people who do not see that the haves and the needs are accurately framed. The ones who see the emptiness are vital to examining a situation.

Discussion of have and need should be done carefully. It's very easy for one confident person to steamroll a discussion that needs to be worked through. If that person's view prevails, it can make it impossible for all involved to have a good sense of what's going on and what they will need to do.

Only once the initial a and the final b have been constructed to the satisfaction (or partial grumbling) of all can the question of how to make an f so that f(a) = b can be taken up.

At this point things can fall apart if anyone pushes their own way as the only way or their own view as final objective truth.

Every process has multiple entangled threads of action. Separating the strands from different views and spinning them together can create a way that works in more dimensions than a single process will. One person's slog is another's dance.

Let the dancer carry the weary, then swap around. My dance is your weary and my weary is that another person's dance, and so on. Done properly, such a discussion will distribute what needs doing to who can do it well.

If the discussion ends up with one or two people doing everything, then take apart the threads so that each person who is actually trying to take part has an opportunity to add what makes their ways work to the threaded shared work.

The separated strands and threaded awareness will also make it easy for people to know whom to go to for help in the parts that are hard for them.

As work progresses, most of the collaboration will likely be in twos and threes working on entangled subaspects and processes. Everyone getting together for other big discussions is unlikely to need daily or even weekly meetings.

It is very important to be aware that people have their own cycles and pacings. One cannot demand that people create to a fixed rhythm. There will be more about this in almost all of Section III.

Contemplation and Conversation

Working together is made of working alone. Entanglement requires what is entangled to be able to act in its own ways relative to the overall ways. The conversations people have also require each mind to be contemplating the whole and their aspects of it.

The problem of how to solve a problem spawns the problem of how to talk to each other about how each person is thinking about the problem.

A bit of conversation may spark the beginning of an idea in one participant. That idea may zip into the back of mind and start a complete tectonic shift which four days later may produce a better solution than what the conversation produced.

Conversations between artists need to have room and respect for, "I'm working on this bit. Can we discuss other parts and I'll come back when I've got it worked out?"

This is, of course, easier when the people involved know each other well. People have external tells related to their internal ways. If you know someone's internal processes, you'll likely be able to figure out that they're working on something. You won't know what it is or what will come of it because it's their mind and every artist is unique in what they come up with.

It's also possible if you know someone well to tell that they've checked out of the conversation and aren't actually taking part. Again, see the history of musical bands and collaborative songwriting.

The better people know each other, the deeper their entanglements and the more possibilities for mutual clarity – but also mutual distress.

People who don't know each other and are put in the same place to work together quickly on something are essentially being given a task that requires a group who already know each other and how they work.

The problem is solvable, but only if you remove the "quickly."

Each of the people involved who are trying to make this work will be doing multiple things simultaneously and all those actions will be entangled together, but not necessarily comprehensibly entangled. Things done include the following:

1. Contemplating the problem before them.
2. Trying to get a sense of the other people involved.
3. Reacting to the method of conversation and the environment. Some people get stressed in physical meetings. Some have trouble with online chats or video calls. No method and environment will be comfortable for everyone involved.
4. Trying to get a feel for other people's approaches.
5. Using whatever methods each person uses for having discussions with people they don't know.
6. Likely not noticing whatever they are doing that interferes with these processes for themselves and for others.

It will take time, care, and effort for each of the people involved to help themselves and each other unthread these things and begin to be able to actually get to work. Each person involved will have their own methods of time, care, and effort and their own difficulties that can get in the way of all of these.

As an evolving group, they will need to not only work out the problems they are working on but work out how to work out those problems. Each and all must be aware that they are solving not just the functionality they are creating but the metafunctionality of their working together.

With care and thoughtfulness, they will be able to craft a function space that will be of use in this and later shared work and play.

Conversation needs to leave space, time, and energy for contemplation. Contemplation needs to accept the "in motion" generated in a conversation, so it is taking part in the shared dance of collaborative art.

Personal Exercises

1. Look at some code you are working on. Examine the functions you've made. See how they look to you using the metaphors of locale, called-in specialist, tool, and river branch. See if any or all of these help you get a better sense of how the

functions work and how you can make them work better. If none of these help, see if some other way of looking at them pops up in your mind in reaction against these metaphors.

2. Try to see what you personally need to make a function seem real to you. Consider processes, associations, interactions, naming conventions, and where it might fit in your personal toolkit. Also feel free to add other considerations that work for you.

Testable Exercises

1. Take any function you've coded. Discuss what it is invoked for. Lay out the purposes of the main body, note any side effects and why it returns what it returns.

2. Take a function that has application to multiple situations. Pseudocode the function for each of these situations. Note the differences, if any, between the code for situations that interest you and those that don't.

3. Create an inheritance structure for an object that has real-life categories and subcategories. Pseudocode at least three methods that will need to be changed as one goes down the taxonomy. These methods should display differences between parent, child, and grandchild objects.

12

Interfaces: Creating Shapes, Flows, and Empty Spaces

Surface as Expression of the Art

Arts rely on the fact that humans can't directly perceive the insides of what's going on around them. Our sense of sight forms an awareness of moving and changing surfaces. Our sense of hearing perceives vibrations radiating through media. Our sense of smell perceives chemicals given off. Our sense of taste involves feeling what's on the surface of what we're eating and then biting and chewing in order to find new surfaces. All the touch senses are surface contact.

Surface as we're using the term does not mean something strictly two-dimensional. Our senses, especially touch and taste, press in and out. We can see through transparencies in order to aggregate our surface vision and hear vibrations coming from inside things. Overall our sense of reality is composed from deep phenomena radiating through surfaces.

Art works this way as well. We build the interior structure of an artwork. We shape the exterior through which the interior meaning and lower layers will express and through which the audience will contact and interact with the art.

The surface is the face of the art. The interface is the entanglement between the art's face and the audience's faces.

Interface/surface as expression of the art involves crafting something that your users will find beautiful and easy to work with.

There are very few general rules of art. Each artist and each audience member is a unique person with unique tastes, interests, and awarenesses. But one good rule of thumb that works in most artworks and applies to all arts is All Emphasis is No Emphasis.

You need to use contrast in your art, not just composition. If everything is loud and nothing quiet then your audience will adjust their perception to the volume (or go away if they can't handle the volume). If everything is the same, nothing can be pointed out.

If you make an interface full of lush 3D moving models, some of which are controls and others are just there to look good, your users will not be able to find what they can actually do.

This does not mean that parts of the surface need to be dull and uninteresting. Rather, it means that finding the interest in the unemphasized parts is something the audience will need to deliberately focus on.

But don't worry about the de-emphasized being neglected. Some of your audience will lovingly explore the de-emphasized. People have fun doing that.

DOI: 10.1201/9781003459866-14

Fanfic writers look at the lives of background characters in stories and shows and make new art from their impressions of them. Enjoyers of sculpture will examine the veins and spots in the marble carved to make a graceful statue. Users will use minor tools for major effects once they get the hang of using the program overall.

And therein lies the vital factor: Emphasize what needs to be perceived and learned in order for the program to be used at all, but make sure other aspects are perceivable and can be worked with.

This problem is illuminated by menu systems.

Menus suffer from a flatness of emphasis. Everything in a menu is equal. The use of submenus does not change that fact.

Menus were invented so that users did not have to memorize command lines and retype things over and over. This, for most users, was a great improvement in interface design.

However, that improvement involved copying the concept of menu from restaurants wherein diners are presented with a listing of possibilities.

Computer menus are organized by kinds of actions, such as edit menus in word processing programs, or kinds of objects to be affected, such as the currently ubiquitous file menu. These arrangements don't necessarily fit the user's work flows. In this situation the user needs to create their own internal emphasis. That is fobbing off artist work on the audience which you should not do.

One solution to this problem is to change emphasis with change of circumstance.

Arts that involve time, such as writing, acting, dance, and music, have methods of changing emphasis as basic techniques. Change of action gives change of emphasis.

Relatively static arts, such as painting and sculpture, rely on users to be able to modify emphasis by changing where they are relative to the art and what part of it they're looking at.

This has more effect than most are aware of. Change of position not only affects what part of something one is looking at but also the relative shading of the object.

Here's a formula I figured out and used in some animation software I wrote quite a while ago. I was trying to figure out the shading effects of lighting and viewpoint. I talked to my wife who had learned a general principle when she studied scientific illustration that the shading depended on the angle between the point of view and the normal vector of what the viewer was looking at. The shading would be 0 (black) if one were at right angles to that normal vector, and 1 (unshaded) if one were looking straight at the normal.

This value is the absolute value of the cosine of the angle between those vectors, which is pretty straightforward to calculate.

$$\text{Cos(angle between v and w)} = v \cdot w / \text{norm}(v)\text{norm}(w).$$

What this means in real life is that as one moves around an object, the apparent shading will change so that what parts of the object are emphasized (brighter) will be easier to examine in detail than parts that are deemphasized (darker).

Programming has a more direct interaction with emphasis than most other arts. What is emphasized is a matter of what persists. What the user interacts with for longer stretches of time or interacts with repeatedly is emphasized. Brief interactions are deemphasized. Everything that persists can be changed either subtly or dramatically as the user interacts with it.

The entire evolving surface interface can and should be seen in terms of what it emphasizes and in what circumstances it emphasizes it. As you build and refine your interface, take a step back and see what is being emphasized, what transitions of emphasis are possible, and how those changes in emphasis affect your user's experience of persistence.

Surface as Mirror

As we've said before, when a person experiences an artwork they entangle their own awareness, knowledge, experience, and ways of perceiving with what they are currently experiencing. The surface of the art acts not just as an expression of the artist's work but also as a mirror of the audience member and their mind relative to the art.

In most arts, awareness of this mirroring can provide the audience member with how the art tastes to them, and perhaps some insight into themselves. How an artwork mirrors a particular person can also be useful information to people who know or wish to know that person better.

Some people will overuse this mirroring as if it were a test of morality or psychology. But it's rarely as simple as that.

Artworks are tools, and mirroring of tool use is complex. The tool in one's hand can become an extension of one's mind in action, or it can feel off and distorted and create a feeling of unease or distress because of disparity between tool and mind. Ease and difficulty with different tools can give someone a sense of how they do what they do based on which ways of use work well and which work badly for them.

Metaphorically, a tool made for someone who thinks in one direction has a shading based on how much of an angle there is between that direction and the directions that work for a particular user. Some tools are bright and clear as if we are looking straight at them. Others are dark and hard to comprehend as if we were looking athwart them, so they are thwarting us.

When an artwork works for a person, the art mirroring can be immersive. People reading stories or watching shows can form shaped echoes of the characters they are seeing and draw those echoes within their own thinking. Their version of the characters will be unlike nearly anyone else's because each such character is entangled with their mirroring.

Fandoms are full of people arguing about the true nature of characters. What they are presenting are the characters that dwell within each of their minds. Different minds have different character immersions, different points of view, different fanfics. All of that is good and fine.

Mirroring with software combines tool-use mirroring and immersive-character mirroring. The feel of using the software and the character it expresses back at the user when actions are taken give an interactive personality to the experience.

Software is not alone in this kind of combined mirroring. Many people have experienced this with vehicles. Cars and boats are often thought about and talked about as a combination of tool and character.

People also sometimes try to make this form of mirroring from their interactions with other people. This is a catastrophic error. People are not as small or simple as tools or characters, and the memories we have of other humans should not be as small and simple as our memories of tools and characters.

Learning about other people has no end and no fixed dimensions. Everything humans do is art and each person's art is unique to and from them, but the artist is not small enough to fit within the body of their work. The artist is always vaster and more multidimensional than any artwork they make. Things about artists can be inferred from their work, but those inferences do not encompass all the possibilities of the artist.

Each person's reflections from the art they experience are unique because each person is a unique audience/user. In considering the reflectivity of your work it is unwise to presume what anyone will project or perceive of themselves in it. They will make of it what they make of it.

How then can we craft reflective aspects?

First, be aware that the users will be reflecting through your art. Make your art as much an expression of your ways of thinking as possible so that your audience will have an opportunity to see themselves in a new mirror and therefore in new ways.

If a thing feels like it belongs in your art, put it in so your audience will have an opportunity to play with it. Where to put it and how to make it fit is the complex part. But do try to do the work to put in what belongs. Refine the element so that a person does not need to share your exact life experience and knowledge to be able to make use of what you are putting in.

Second, remove stumbling blocks. If you put in something that is easy to trip over or confuse, then clarify and emphasize the ways the thing can be used and turned and examined.

Third, never presume on a reaction. If you construct a multistep process that your users will need to follow, do not presume what they will like or dislike, nor what will be simple or what complex for them. Make your processes broad enough so that people won't stumble over the need to do what is obvious to you. Slow down the process enough to make the ways and the various branches of those ways visible to your users.

Most actions that change what persists can be expressed in a formulaic way as "Use this tool with this technique on this material to transform this artwork."

But this general description has four very broad parameters, each from its own spaces of tools, techniques, materials, and artworks, as well as having two very large function spaces of uses and transforms.

Three of these – tool, material, transform – can be supplied by your software. Tools are functions. Materials are data structures. Transform is code being run.

The software also needs to be able to hold the dictum of the artwork so that the user can perceive it well enough to be able to work on it.

Use, technique, and res of artwork are reflected from the audience's mind. Those reflections depend on what you have supplied as a surface of reflection.

You and your audience are each operating in large personal spaces. The program should be coded so that you and your audience are each trying to help the other contribute their own aspects. If this is done well enough on both sides then the entangled awareness and action of changing persistence can flow in ways that are useful and enjoyable for the user to do and for you to have made possible.

The user will get the direct feedback of your software, and hopefully you will get the indirect feedback of hearing from people who benefit from your software.

The Shape of Output

The evolution of the output of a program depends on three basic questions:

1. What in the output is static regardless of user input?
2. What in the output changes regardless of user input?
3. What in the output changes depending on user input?

Most programs that are made to be work tools use a primarily static structure with elements that change only depending on user input. The word processor I am writing this book on is a good example. This structure minimizes distractions and makes my own actions and their effects quite visible.

Many games present an evolving output with changes happening without regard to user action. Some of these are actual game elements that the user will need to respond to, but others are ongoing effects to give a sense of an active environment such as changing lighting based on day night cycles or clouds moving or little animations of background objects and so on.

Programs that play videos or audio will have the media play unless the user stops them and keep most of the rest of the program static so as to not distract from what the user would like to attend to.

As these possibilities illustrate, the answers to the three questions depend primarily on what the program is created to do and how the user will need to react to what the program does.

These practical questions and their answers all flow into the more artistic question of what output shapes can you possibly create.

The answer to this begins with hardware.

What devices is the program going to run on and what can the device itself do in terms of shapes? Indeed, what kinds of real-world phenomena can the device generate and therefore shape?

I'm writing this in a time when visual and auditory output are pretty easy to create, tactile is being worked on, and other senses are purely experimental. I can reliably shape appearance on screens and sounds through speakers. Beyond that things are, for the nonce, nebulous.

Visual output has had the longest evolution and the easiest path to evolve along because our sense of sight is conveniently pixilated. Many aspects of two-dimensional and some three-dimensional art can be directly translated into computer representations. This field has a large array of tools to play with and you should be able to find such tools to help you create the visual possibilities that work for you.

If you don't find such tools, then you may need to code your own. In either case, the visual shape of your software is something you will need to work out, create, test, and modify so that the mirroring and immersion of your visual elements are close enough to the res in the back of your mind.

The difficulties here are the difficulties of all visual art. Current design tools let one generate okay surfaces. If you need to go further into depth to bring your work to life then you will either need to develop your own visual art skills or work with a visual artist as consultant.

Our other senses aren't naturally digital. Hearing is so much an analog sense that it uses biological drums and horns to gather in the sounds.

Our ability to make effective digital music and voices depends on the combination of two interesting bits of mathematics.

We talked briefly earlier about how Taylor–MacLauren series makes it possible for computers to reduce sophisticated functions to the addition, subtraction, multiplication, and division that computers can actually do.

Going from our analog sense of hearing to digitally recorded and generated sound could therefore be accomplished if sound could be represented using the kind of mathematical functions that can have Taylor–MacLauren series.

Fortunately, this problem was solved long before computers were invented. The mathematician Joseph Fourier, who lived in the late 18th-early 19th centuries, figured out how to represent a complex wave structure (such as sound, for example) as a sum of series of various kinds of functions.

The most common Fourier series uses the sine and/or cosine functions. Such a Fourier series breaks a complicated sound up into a sum of pure tone waves, each of which looks like amplitude * sine(frequency * time + phase shift). Amplitude gives volume and frequency gives tone.

There are also more complicated Fourier series, such as the series of Bessel functions, which do a better job representing drum sounds than the sine and cosine series.

A mathematical method of transformation from analog to digital sound is to represent the analog sound as a Fourier series, then use the Taylor–MacLauren series representation of each of the functions in the Fourier series to turn the sum of sines into a sum of polynomials, then have the computer do the add–subtract–multiply–divide to generate results that are close enough for human hearing, then put the output through another conveniently lying-around technology: Speakers.

Each sense other than hearing and sight is more complicated than this and requires more specialized hardware. I refuse to predict what will be developed when, because as an SF writer I'm aware of how bad the track record of tech prediction is. On the other hand, I don't want to restrict this discussion to hardware that exists at the time of writing. So I'm going to stick with a general discussion rather than focus on particulars that could be on their way out while I'm typing this.

So, what can we generally say about hardware as presentation space? Users will need the means to perceive the output the hardware generates, which will create different habits among different users. They will have to be able to focus on the output in ways that will make other sources of the same kind of output either blend with or distract from the output/the output will distract from those other sources. For example, the smells your hardware give off will either draw in the user's attention or distract them from the soup they're eating.

When coding, be aware of how much attention your software will require from your users. Are you requiring 100% of their attention? Will they need isolation from outside stimuli? Or can your output be perceived while someone is conversing or watching something else or dining? Can they easily shift from your software to another program running on the same or another device?

It's easy to work on the assumption that a person should be attending to your work and only your work, but it is an unwise assumption.

By the same token, it's possible to make software that is meant to be run in the background of someone's life and not benefit from overmuch attention.

Before constructing output, consider the range of real-world life attention that needs to be given to your program and the hardware it is running on.

Okay. We've got the user and the hardware and a sense of what the hardware can do. So let's pass inward to the persistent output that the hardware is generating for the partially attendant user.

It is current practice to not have multiple sources of sound output actually outputting sound to the user at one time because sounds blend by adding Fourier series and can interfere with each other. It does not take too many simultaneous sounds to produce the too-much-information we experience as noise. So one sound source at a time is good design practice.

Visually, if we look at a screen, we see software shapes nested within each other. There is an outermost presence of the software that regulates the running of all other software on this hardware and there are the diverse shapes of the individual programs running. Within each program there are visual shapes for controls and various displays.

Our sense of sight evolved to compose multiple sights together into a single evolving awareness. It is easier still if we can focus on each thing when we need to focus on it.

The most common physical methods of focusing are shifting our point of view so we're looking directly at the thing we're focusing on, and moving closer to it (or moving it closer to us).

This does not work so well in the visual environment of screens because of limited physical space.

The most common solution is to be able to move what is where upon the screen and to be able to change the size of what we're looking at. These produce the same front, center, and large vision effects as turning and getting close to.

One oft-neglected consideration is the texture of the output. How sharp does it feel? How fuzzy? How much does one part of the output blend with others?

Most visual, auditory, and tactile arts are aware of texture as a basic consideration. As they use materials made of information, they can take advantage of the fundamental fuzziness of reality.

In computer output, however, we are working with data and have to handcraft output fuzziness. Overly-sharp edges in sight, sound, smell, taste and touch have an intensity that may not be the effects you are after. Fuzziness can be comforting, if it does not distract from ability to do what is needed. Fuzz things up as you need to give your output the feel you wish the user to experience.

The Flow of Input

Just to clarify, we are using input to mean direct input from a user that the user is aware of inputting.

There are a number of languages that make little or no distinction between a live data stream from, say, a video monitor, a data stream of the contents of a web site, the readouts of a heart monitor, or a data stream from a user typing.

These languages are focused on input as a way that data enters the computer. But we need to also be concerned with input as part of the audience experience.

From our perspective only the user actions are input. The rest of the above examples are storage/retrieval.

There are diverse kinds of input. Just as with output they are very hardware dependent.

In terms of general design and user experience we can examine input in four aspects: Consequence, knowledge, timing, and commitment.

Consequence is what the software does with the input. This can vary from user actions that do nothing to the user throws the switch that brings the monster to life to the user pushes the button and destroys the world.

Consequence in a nutshell is what function is invoked when the user does their input action.

One crucial aspect to consequence is what feedback the user receives in the form of output when they do a particular input action. Action without feedback does not feel like action. It's very easy for users to get annoyed, frustrated and ragey about actions that seem to disappear into the computer without responding in any fashion.

Knowledge is the consideration of what the user needs to know and understand in order to meaningfully input anything. A user presented with a set of buttons with obscure symbols has no idea what will happen if they press any of them.

This is a common situation in certain kinds of puzzle games. Such games incorporate the idea that users will learn their own capabilities by trial and error.

Unfortunately, this idea is not confined to puzzle games. It's common in software tools for certain specialty fields to presume that a user has enough technical knowledge of the field to understand what a particular label will mean and know therefore what the control is supposed to do.

This shows that there are two dimensions of knowledge here, knowledge of the field and knowledge of the software itself.

It is a good idea to cover some of each aspect of knowledge in documentation, because not everybody skilled in a particular field knows the same things and not everyone is taught the same techniques and symbols. Rather than presume, it's a good idea to be explicit about how a control implements a tool, not just that it does so.

Except specifically in puzzle games, it's not a great idea to have your controls be obscure.

Timing is the consideration of whether consequences depend on when the user does the input. This largely depends on the timing of other things in the program.

Real-time games are strongly timing-dependent. Combat in such games often requires that players learn the cycles of an opponent's actions and learn when to run and when to strike.

Timing can also make a large difference in less dramatic programs. Pausing a video at a particular spot in a scene (or more extremely, at a specific frame) can require precise timing, especially if the video software does not have an option to step frame-by-frame or second-by-second.

At the opposite extreme, many programs simply wait for user actions and process them the same way regardless of when they happen. This word processor has that sort of timing. It makes no difference to the text when I type these words and there is no sign in the text itself as to whether I got up and danced a jig between the words "danced" and "a jig".

I didn't, by the way. Jigs are not a dance I know how to do.

Timing should be engineered for the needs of the user and the ways of the program. Every demand of precision in timing is an exclusion of people who do not easily move at the demanded pace.

Commitment is the complexity, attention, and focus needed for the action taken to be able to do the input. Pushing a single button is usually a small commitment unless the button is very small and hard to find.

One push is one user action to produce one software action.

But even a button push can have commitment precursors. Does the user have to set a number of values or move objects around on a screen and then perform a single button push? How many individual steps and how much focus for each step is necessary to make a single change?

Commitment also measures how much of the user's artistic abilities is involved in the input. Drawing a picture takes a lot more commitment than uploading a picture file. Writing takes more commitment than quoting someone else. Commitment is a question both of attention and ability.

Consequence, knowledge, timing, and commitment are ways to measure the flow of input as the user experiences it. They are what you are invoking from your users when you ask for input.

Consideration of these four aspects will make it easier to figure out what kind of input method you should employ. You may be able to find a standard input control or bundle of standard input controls that do what you need to do, or you may need to make your own control to give your user the input process they will need for your software.

Input is a flow, not an individual action. It is the user's ability to affect the program and its contents. There is a cycle of input, consequence, output, consideration of effect and going on to the next input. This cycle involves the user's mind and consideration, but it also must reconcile with the timing of the input.

Based on the user and the program's abilities to mutually adapt, the user may develop rhythms and habits that flow well with the program. If these do not develop, the user is likely to be disaffected and give up on the program or keep using it while hating and raging internally.

Each user will develop their own rhythms and habits relative to how they experience the consequence, timing, knowledge, and commitment of the program. In short, the user will react to how you are invoking them.

Unlike the computer, users can experience frustration, confusion, and rage. They can also experience happiness, interest, inspiration, and artistic fervor if their ways work with the flows of your software.

It is therefore a good idea to give the program itself a level of adaptability and let there be more than one way for a user to do something. Make it possible for the user's ways to take up the ways of the program and bring them together.

Space, Time, and Possibility

Art changes the possibilities of people's lives.

Sometimes it does this directly, by giving people capabilities they did not have or ways to make use of capabilities they do have. Sometimes it does so indirectly, by giving them insights and perspectives they did not have access to before interacting with the art.

Some art does this by becoming part of their lives. People give space and time to the art and the possibilities from the art grow and become part of how they live.

Such art can compose itself with possibilities already present in their lives, or it can create new possibilities in their minds and lives, or it can make connections with other people possible so that the connections can increase possibilities for all the people involved.

In all of these cases the art must fit the person's ways and life. For software the job of fit exists in the interface, in the input and output. What you design and what they use will ask for a commitment of time and space, energy and thought. From these it can produce otherwise unknown possibilities, both for the user and from them.

The purpose of the interface is to open up the possibilities within your software to your users. If you find yourself losing track of that, getting lost in one feature or another, one bit of shininess or another, or one bit of overelaboration or one omitted way people might use it or another, then try returning to this awareness and these questions.

How does this help the user open up the possibilities?

What does it require of the user in space, time, energy, and thought in order to open those possibilities?

Personal Exercises

1. Take a static artwork and move around it (or move it around in your hands) and get a sense of how its emphasis changes based on where you are relative to it and how its different aspects are revealed and concealed based on your actions.

2. Take a dynamic artwork (such as a book, video, or audio recording) and read, watch or listen to a chunk of it. See if you can discern what is being done in it to affect emphasis in your experience of it.

3. After doing testable exercise 2, see if you can get a sense of the mirroring aspects of the program. How does your use of it reflect the way it interacts with your life as you live it.

Testable Exercises

1. Take a program you use a fair amount. Record and describe what the program emphasizes and deemphasizes.

2. Using the same program as in exercise 1, make note of how your workflow is facilitated by and made harder by what you made note of in exercise 1.

13

Practice, Experimentation, and Playing with Possibility

Recurrent Tasks and the Development of Practices

The metaprocess that we've used for developing functions might look like it can also be applied to human activities:

What do I/we/they have?

How do I/we/they get/make what we need from what we have?

What do I/we/they need?

But this structure really only works when working in a data environment where particular things are scarce and all processes need to be constructed. It works best if persistence is controllable.

None of these conditions apply to humans and human practices.

Humans evolved in an information-dense evolving probabilistic environment where entangled interactive activities are going on all around and within each individual thing or being.

At first glance such an environment might seem too confusing for anything to develop that could work more than once. But this is an inaccurate perspective on probability and entanglement.

The likelihood of an outcome to a probabilistic event depends on the circumstances of that event.

If circumstances recur, then likely outcomes are likely to recur.

Evolution therefore will select for what can make use of recurrent outcomes.

But why would circumstances recur?

For this we go back to the roots of chaos theory. What happens when the same process (say, gravity) is iterated over and over again on a set of initial conditions?

Consider the relatively early universe.

Not really early – we're not talking about all the wild stuff in the first picoseconds or even the first few minutes.

Consider the universe early enough that we've got hot baryonic matter nuclei (mostly hydrogen, some helium, a dash of lithium and beryllium) and a lot of electrons flying around, and lots and lots of neutrinos.

DOI: 10.1201/9781003459866-15

Note that the model we're using doesn't take dark matter or dark energy into account, although they have effects on the resulting arrangements of baryonic matter.

At this time the distribution of matter is uneven and the universe is still rapidly expanding. Where matter is relatively dense it pulls together through gravity. Where it's relatively rare the matter gets pulled into the dense areas.

In *The Fractal Geometry of Nature*, Benoit Mandelbrot called this process "curdling". It's an evocative term, if not fully accurate.

The curdling process fractalizes. Denser areas within the dense areas pull together even more. The emptier areas empty even more.

Eventually we get metastructures of what will become galactic clusters and what will become galaxies.

At a very small scale (relative to the entire universe), the curdling leads to small, very hot gas clouds of mostly hydrogen pulling together into star-sized lumps of gas.

If there were no other processes going on the gas would eventually crunch into black holes.

But two other processes kick in as these nuclei get close together.

The second process is uncertainty pressure and the Pauli exclusion principle, which together make it difficult for lots of fermions (including baryons) to occupy the same position and energy state.

The first process is fusion.

Those hot hydrogen and helium nuclei which have been accelerating due to gravity slam into each other and fuse, becoming hotter and faster and no longer only moving toward each other. In a dense enough space fusion has chain reactions, causing more fusion.

Chain reactions are probabilistic phenomena. They become more and more likely depending on how much hot smashy nuclear stuff is happening in a large dense area.

When the likelihood of chain reaction gets high enough, we get stellar ignition.

This process recurs in all the denser regions of denser metaregions because the underlying physical laws are the same everywhere. Stars, as long as they have fuel for fusion, acquire a stability because the inward force of gravity is countered by the randomly directed motion and energy released in fusion, as well as the uncertainty pressure in the core.

This stability only lasts for millions or billions of years. It can be seen as just a blip in the long-term gravitational evolution of the universe.

But it is stable long enough for other shorter-term processes that treat the stability as an environmental fact to develop.

It's worth looking at the pairing going on here. A large-scale phenomenon like gravity determines an overall contractive condition, which is then countered by a small-scale, in-motion, energetic phenomenon like fusion, which presses outward against the contraction.

This countering is not a simple $A - B = 0$ kind of thing. It has more fuzziness and random possibility. $A - B$ is close enough to 0 for other phenomena to spark up and start doing things.

Similar countervailing effects happen in planetary formation, in the determination of orbits, and in the evolution of climates, life, and things like that.

Here we are steering between mountains of books.

We will, however, make a brief dive into a kind of recurrence we are much more familiar with and which serves as a good example of what we evolved to deal with: Day/night and the turning of seasons.

The rotation of the Earth about its axis and its axial tilt combined with the particular elliptical shape of our orbit around the sun shape the cycles of light and darkness as well as the basic flow of energy into the system of our planet. This combines with how water and air change when warmed and cooled and the interior processes of the Earth to create the basic conditions of our varied climates.

The recurrence of day and night and the turning of the year create interesting evolutionary results. Evolution happens in response to past and present pressures. But it is possible for creatures to evolve things that will benefit them in the future because that future looks like the past.

Evolution does not anticipate, but it does select for what will work in recurrent circumstances. A characteristic that would not be a survival characteristic if day were eternal or if water never froze in winter can be selected for because the organism needs to survive and can make use of day-night cycles or seasons in temperate zones.

Past pressures create future benefit if local time has cyclic and recurrent phenomena.

We will now sail around the entirety of earthly evolution and jump into our own minds.

Human learning involves developing the capability to deal with the future based on resemblance to the past. We can anticipate and can consider not just the facts of night and winter but the question what might we do when tonight or next winter comes.

Our minds seem to have evolved to anticipate, so much so that we project anticipation on to other things. People imagine that evolution must have forethought because we look ahead when selecting what we will do, where evolution works by recurrence.

Our forethought is a survival characteristic because it evolved as a means of making use of recurrence. We anthropomorphize evolution by presuming that the only way to have a characteristic that depends on the future to work is to anticipate it.

This has been a drawback in understanding evolution, although we're learning better.

Forethought gives a way to think of human learning as the ongoing development of an individual's capabilities to deal with kinds of recurrent phenomena. This is unlike coding a function because both the individual and the phenomenon will be different each time the recurrence happens. Days are alike in overarching event but not in what happens in them. We can't just code a formula for daily activities and run them automatically.

Because, again, we are dealing with an information-dense and possibility-dense real world environment, not a data-sparse, possibility-narrow computer environment.

Knowledge, Understanding, and Capability

There are three aspects to this kind of recurrent past anticipating the future learning: knowledge, understanding and capability.

Knowledge is awareness of what is happening in the recurrence.

Understanding is awareness of how and why the recurrence is happening and which ways it can evolve.

Capability is the ability to do something with or about the recurrence.

All three of these are entangled. To be unnecessarily fancy and because I like the word, let's call this entanglement of knowledge, understanding, and capability praxis.

There is a tendency to name a specific praxis based on capability, such as calling all aspects of programming by the name programming, which is the capability of writing code.

This can lead to circumstances where a praxis is developed but is nameless because the capability is not what matters to it.

For example, it is quite possible to learn the history of painting and understand how painting is done because one is developing the capability to help a painter with their work without developing the ability to paint. That praxis has no name, but it definitely exists.

There is a similar praxis in writing which has a name: Editing. A skilled editor doesn't need to be a capable writer in order to help writers make their writing better.

The fact that editor is a named capability and the equivalent for painting is nameless is purely an artifact of the evolution of the corresponding arts. Writing happened to evolve to go through the process of publishing. Publishers hired editors to help the writing meet their standards and/or needs. Painters had no equivalent of publishing (Art galleries are showrooms, which is not the same thing).

The point here is that a praxis need not be known or talked about or named to be useful. You should develop the practices you have uses for, whether or not they have names.

Knowledge and understanding also have name problems. Every area of human endeavor is a complicated evolving fractal and each person who knows some of it will have their own personal interplay of it.

A person who, for example, has read Aristotle and the medieval scholastics will have an idea of what logic is. But that concept of logic is archaic compared to a person who has studied 19th–21st century logic. And someone who has studied that more recent logic will have different attitudes toward the field if they study from a mathematical versus philosophical perspective.

All three of these people might say with some accuracy that they "know logic."

Of the triad aspects of praxis, understanding is the most difficult to put into words because it is back-of-mind centered.

Understanding is the capability to work with the res the praxis operates on to see and feel how that res interacts with other aspects of the praxis and the other praxes of the mind practicing it.

Understanding reveals what parts of the res are recurrent and what parts are unique to circumstances. The understanding of a praxis guides the mind in how it evolves res and dictum relative to recurrence and uniqueness.

Understanding is always grounded in how the particular mind understands things, which is itself a combination of recurrence and uniqueness.

This grounding is a combination of innate awareness of what feels like it works for you, evolved experimentation, and seeking out possibilities. We'll talk more about this later in this chapter.

The main thing about understanding is to not try to pin it down or insist that it has to work according to somebody-or-other's theory. Each mind needs to make and find its own ways.

Capability is the most front-of-mind aspect of praxis. It evolves and grows with practice.

Exactly which practices work for a particular person depends on that person and how they practice and learn.

Capability cannot be learned without trying things and examining results. Most teaching involves the setting of exercises to give practice. But in many cases, those exercises presume that all students share a single rhythm of practice and learning. This can lead to errors in teaching and in examining how students are doing.

Consider a person who can do, say, ten of a certain kind of exercise extremely well and then their mind starts processing the effects of doing that exercise and in effect shuts down.

But suppose they have twenty more exercises to do and do not have time to wait for their mind to recover.

Those later twenty exercises will come out badly. It looks by the metric of percentage of work done that they are either not learning or not trying, although tomorrow or the day after they will be capable of doing all of that kind of exercise.

The rhythm of the teaching here does not fit the rhythm of the learning. The student may become discouraged and think they cannot learn to do this.

A wise teacher who has time to analyze thoroughly would notice the pattern of early correct answers followed by later incorrect ones and might seek to correct their teaching and testing methods for that and any other similar students.

This is one of the areas where art teaching has much to give to other kinds of teaching. Good art teachers are aware of the individuality of their students in terms of what kind of practices help them. The teachers need to have this capability because good art teaching is not looking for uniform results among its students, it's looking for each student to develop their own capabilities.

Teaching capability means helping each student develop their own way to produce the results that are needed in order for them to learn to take part in the art being taught.

Praxis and Languages

The ways various praxes are talked about can change how easy or difficult it is to consider, develop, and evolve which praxes a particular person needs.

Consider the following two statements:

Speaker 1: Method A works.

Speaker 2: Method B works for me.

In everyday usage this would be heard as the first speaker affirming a general truth and the second speaker trying to create an exception in that general truth.

But the first statement does not even affirm that Method A works for the speaker, let alone for anyone else. It is a vague, dramatic claim. The second statement is a specific operational claim about the relationship between the method and the speaker.

It would be unwise to entrust Speaker 1 with a task that would use Method A because they haven't said they can do it. It's often wiser to entrust Speaker 2 to do a task that Method B works for because they're actually claiming capability.

There are also problems with comparatives and superlatives.

"Method A is better than Method B" implies a single method of ordering. It does not ask in what way it is better or for whom or what.

For example, in the development of linear algebra, a number of subtle methods to invert a matrix were invented because the brute-force calculation method is messy and liable to errors when humans do it. But the brute-force calculation is usually the best method for a computer to do.

Which method a human should use if they have to do it by hand depends on what works well for that human.

If they have a computer they can use the method of having the computer do it by brute force.

"Method A is better than Method B because one needs only to do these tasks to do method A, but one would need to do these other more complicated tasks for Method B" is a better way to express it because someone could respond to the claim that Method A must be better with, "I can't do some of those method A tasks at all, but I'm well practiced in the Method B ones."

Superlatives are the worst.

> All people should use Method A, the one true best method. No others should be tried. We have found the best practice. Let all follow our way or be cast out of this field of endeavor.

Short declarative sentences coupled with exhortations make certain praxes into causes that people champion and seek to spread. Complex relativized statements give perspective and context and make it easier for people to determine for themselves.

Before getting down to finding, adapting, and making what praxes work for you, it can be needful to clear away the language one is using to think about praxis.

Inspecting ones' assumptions and clearing away the ones that interfere with finding and developing one's ways can be good praxis.

There are many ways to do this.

Some praxes are preparatory. These are done before venturing forth into work or play. For those praxes, one should inspect and clear away assumptions before considering what one will do.

Other clearing praxes are done while trying to do things. For those praxes, if an assumption pops up (sometimes as a voice or memory) saying that one should do the one and only thing one could do, it is generally wise to take a moment and consider if that assertion is really the only possibility. Look at the situation from multiple perspectives and see if there are actually other ways to do it which might work better in this particular situation.

A third set of clearing praxes are done after the fact, reassessing what one did and what worked and what didn't. Reshape the practice if needed to accommodate the new information. Do these because recurrence recurs. That's what praxes are about. This reassessment and rebuilding is often a back of mind process. It's one of the things going on when minds are crashing after doing work.

All of these clearing and corrective practices are subject to complications. It's good to examine the complications rather than try to simplify them.

One major complication is that it is possible for a mind to be fully aware of what can work for it at the moment so that the clear path can look narrow and the distractions from that path can look like broadening of possibility. In this case, the seemingly narrow path arises from the understanding that this way is likely to work in this situation. If the path did not work out then one can examine after the fact and consider alternatives for next time.

On the other hand, if the mind lacks a clear path it can be tempted to grab onto something that claims to be a clear path. There are many things advertised with the feeling of certainty and ease. Feelings like that do not necessarily correlate with reality and advertisement is not a thing to rely upon.

Learning the difference between back-of-mind awareness of what to do and front-of-mind asserted certainty can take a long time and a lot of practicing the process of learning and crafting what praxes work for you.

This can be done by a combination of learning from others, adapting what you learn, and when needful creating new praxes from your own understanding.

Learning and Adapting Practices from Others

Most of human education consists of being taught practices and having to adapt them on the fly to how one does things.

Cultures, families and schools teach what they think people need to learn for their time and place.

This process varies in its flexibility depending on the attitudes and possibilities of the culture, the family, and the schools available, not to mention the teachers within the schools. People emerge from this process with a variety of praxes. Some work for them. Some they can make work but aren't good for them. Some praxes they can't do, but they may not have been presented with any alternatives.

Depending on teaching methods, the learner may be helped to find the praxes that work for them or required to learn by specific praxes that are presumed to be the only way a thing can be learned or done.

Praxes one learns early should always be reviewed later in life to see what works for one, what needs adaptation, and what needs chucking because it just doesn't work for the way one thinks, works, and plays.

People vary as to when in life their own methods of learning are well enough developed for them to determine what they need to learn and from whom they will try to learn it.

Every teacher and every written, recorded, or coded source of teaching varies in how broad a range of methods of learning they can accommodate. Other factors such as class size, teaching load, and strictness of curriculum can make it easier or harder for teachers to teach this particular set of students rather than a hypothetical general set of students.

From the student's perspective, learning from these different kinds of sources depends on how well they can fit their personal learning methods to someone's teaching methods.

A critical part of this is the matter of forethought. If a subject has a fairly strict dependency of future knowledge, understanding and capability on past knowledge, understanding, and capability a teacher will have an easier time foreseeing what a student needs to master before moving on.

For example, there are regions of math learning (e.g. the curriculum from arithmetic to algebra) wherein it is pretty clear what a student needs to be able to do fairly quickly and competently before advancing to the next subject. Later math becomes more diverse and personal thinking dependent because each person does proofs their own way.

But in learning an art there is more than one way to reach competence and more than one form of competence. The ability to improvise jazz in a small group of musicians and the ability to play second violin in a symphony orchestra are both forms of musical competence, each with its own awareness of music. The ability to compose music without being able to play any instrument is another form of musical competence.

A music teacher who can only foresee one of these paths is likely being too narrow in their awareness of their students. A student learning music may or may not have an idea of what future they might be seeking. They are likely to pick up clues based on what skills and praxes they are interested in.

One needs to be careful about this distinction. The beginning praxes in nearly every field are more like A-B-C, 1-2-3, do-re-mi than like story writing, mathematical proofs and playing symphonies.

The dullness of early praxis is regrettably legendary, and the gap between where one starts and where one might hope to go is the stuff of many a weary childhood.

But praxis is built for recurrence and therefore needs to handle repetition. Each mind needs to find how it does that and which repetitive tasks it can master and which don't work for it. A great deal of self-awareness can be hidden away in childhood memories of what came easily and what eluded one's ability to learn.

In any case, learning praxis from a source usually requires an entanglement of foresight between the source and the student. The student may be learning in order to accomplish some particular thing that exists in nebulous res in their minds, or to be able to take part in an aspect of life that interests them, or because they want to see and understand more about a field that fascinates them.

A teacher may foresee their students making use of their teachings in everyday life, or joining the academia of that field and working to teach and advance it, or hoping that the student will follow this learning for the sake of enjoyment.

If the foreseen future of the teacher or other source of learning and the foreseeing hope of the student can accommodate each other, then the student should be able to make their own way along a path that runs near to the way the teacher or other source is following.

Students' learning ways vary in timing and commitment and carefulness and many other dimensions, but they tend to recur in basic methodologies.

Most students are explicitly or implicitly asking for one or more of the following:

1. Show me how to do X, because doing X is what I'm trying to do.
2. Show me the materials and techniques of this art so I can put them together and make what I will make.
3. Show me what has been done before and how it was done so I can see what has not been done.
4. Show me what has been done so I can see how to improve those ways.
5. Show me what has been done so I can exemplify those ways.
6. Show me how to explore what is around in this field so I can expand the field of my exploration.
7. Show me what I need to learn to tackle this one specific task that looms in my mind.
8. Show me how to understand this field well enough to teach it to others.
9. Leave me alone to play with this stuff, then make suggestions of ways to improve what I'm trying to do.

This list is by no means exhaustive.

In the process of being shown the praxis, students will be given one or more perspectives on each aspect of the praxis. Some will find interest in the knowledge, some in the understanding, some in the exercises of capability. Many will find some combination of these in particular applications of the praxis.

It's a good idea to find what interests you most because an effective group of learning practices involves using what interests you to help garner the aspects that you're less enthused by.

Knowledge can be organized in a way to illuminate understanding and see the benefit of competency.

Understanding can point toward why aspects of knowledge and competency are needed.

Competency can seek for greater competency, which requires more knowledge and understanding to be grasped and used.

Once one has a sense of how one adapts learning to oneself and oneself to learning, one will be able to make one's way through a wide variety of fields. One will also be able to get a sense of the methods teachers, books, and programs are using to impart knowledge, understanding and competency, and whether those ways work for one.

Inventing and Creating Practices

There are two entangled reasons to create new practices. The first is if you find that no way you've learned to do something you need to do works for you. The second is so that you become able to do something that you haven't seen anyone else do at all.

There is a tendency to treat invention and creation as mysterious and unknowable. This is in a sense true, in that they cannot be forced or controlled and they're very back-of-mind activities. And yet a person can learn how to be more capable of creating their own praxes.

The idea that invention as a rare thing done only by the mysteriously talented is a more extreme form of the attitude we addressed at the beginning of the book, that only a small set of human works are really art.

Invention and creation go on in the human mind all the time.

Often the results they produce are only of use to the maker. Figuring out how you would enjoy eating a particular meal is an invention of a praxis. It's not something necessarily of use to anyone else. It might not even be of use to you beyond this particular meal.

We're going to focus on invention with longer-term benefit which is therefore worth the time and commitment to create. Praxis, after all, is about recurrent actions.

We'll be using the framework of knowledge, understanding, and competence to give us approaches to the act of invention.

Please note: We hope this will work for most of you, but nothing works for everyone. If this framework only somewhat works for you, gather what you can from it and reframe that in your own ways. If it's actively disruptive to your thinking, then try to see it as a framework that may work for other people and their works and try using it as a lens to see what they are doing and have done. Then consider how you work in contrast to that.

The process of invention usually begins with a sense of lack or of something just plain not working. This is the same beginning as res work because it is res work. So we're going to delve into res work from a knowledge-understanding-competence perspective.

A given mind will usually have one or two of the three aspects of praxis that it does better than the others. These often correspond to what is enjoyed. Some people more enjoy learning the knowledge of things, others the understanding, and others the capability. Use what you do well to make it easier to develop what you have more trouble with.

There are six possibilities for how to use what you do well in order to fill in what you don't do so well.

Three of them take two aspects to generate one:

Understanding and competence develop knowledge.

Understanding and knowledge develop competence.

Knowledge and competence develop understanding.

The other three take one aspect and generate two:

> Knowledge develops understanding and competence.
> Understanding develops knowledge and competence.
> Competence develops understanding and knowledge.

The use-two-to-make-one methods all work together and form the structure of trial and error.

Knowledge and Competence Develop Understanding

This is hypothesis formation. Past knowledge combined with competence in current capabilities makes it possible to consider next steps that might work or hidden previous steps that account for the current situation. From these a hypothetical understanding can be formed.

Understanding and Competence Develop Knowledge

This is experimentation and recording of results. Using one's understanding of how things might work according to the hypothesis, one tests it carefully and competently and observes the results. The whole process can be recorded and examined to see how well the hypothesis is working, which gives knowledge of what happened.

Understanding and Knowledge Develop Competence

This is error correction, wherein the results of trying are examined relative to understanding to see whether the method needs modification or replacement with some other hypothesis entirely.

Different minds will vary in which steps in this cycle they work best with.

Furthermore, the cycle as given here is not the only order in which these can be done. One can start by simply doing something and come out with a surprising result, make note of what happened and wonder why it happened this way. One might then form a hypothesis and enter the standard cycle or one might keep playing with possibilities to get a wider field of data so as to be able to step back and form a wider hypothesis.

Alternately, one might focus in and make small changes in a process in order to fine-tune the effect in order to develop a competence without ever forming a hypothesis about why it works. This will develop competence with a narrower understanding.

Competence is often coupled with a sense of proportion, how much of one material to use relative to how much of another, how to tell when something is in the state you need it to move on, and so on. Practical tests and feelings are the most commonly used forms of understanding. Anyone competent in an art has a number of those practical tests and feelings as part of the understanding of the art.

The practices that use two to make the third overlap easily. For example, if you use competence and knowledge to craft understanding, you can then switch to using knowledge and the new understanding to update competence. Finding which way to arrange these for yourself is a useful metapractice and will over time help you develop how you develop new ways.

The practices that use one aspect to get at two are broader and don't fit together as simply. They are much more subject to the particular ways of each mind.

Knowledge Develops Understanding and Competence

As we said before, human memory is not separate from human imagination. Because of this, our knowledge has the feature of being easy to play with and to connect to things, but also the bug of not being accurate to reality. It's good, therefore, if one has access to honest, testable outside sources of correction for knowledge, as well as interesting and honest outside sources and experiences to increase knowledge.

Minds organize knowledge idiosyncratically, but it's pretty clear that three kinds of association structures are common: Geography-architecture, characters, and events.

Geographic knowledge is so common that it's embedded in vast swathes of everyday practice. Just the instruction, "Go to the grocery store" implies a complex knowledge of one's environment and how to get around it. Before reading and writing became widespread, there were practices to aid memory that primarily relied on geography and architecture to act as memoria. Together these were called the art of memory.

Geographic knowledge develops understanding and competence through exploration and examination of what is where and how it works where it is.

Geographic competence often looks like finding the shortest or most interesting path between abstract places.

Geographic understanding looks like learning the interconnections and hidden processes linking different regions.

People whose knowledge is not geography-based often struggle with the presumptions and expectations formed by those who do have geographic memory. Such people can get more easily turned around and confused as to where they are or where they are going and need other information to help them get where they need to.

Character-based knowledge uses memories of people or characters in stories as repositories of knowledge and as advisors or object lessons in what can be done with that knowledge.

Stories are a common method of teaching knowledge through characters.

Character memories are accessed either first person (imagining oneself as the character), second person (discussion), or third person (witnessing what the character would or might do in various circumstances).

Character understanding and competence look like characters in action or characters explaining action.

Character memory can be tricky to deal with because people will imagine themselves doing things or recall themselves as having done certain things and think that their imaginary selves are the same as who they are as a person now. But all characters are imaginary, regardless of whether we think they are us or not.

A related problem arises in remembering other people. Our knowledge/imagination of someone else does not necessarily correlate well with the real person. It takes a lot of entanglement and interaction to have an accurate memory of even a few aspects of another human being, let alone a full person.

This can get worse the longer it has been since one interacted with another person. The internal character memory of another person can and will evolve in one's imagination and will diverge from the lived experience of the other human.

Imaginary characters can have any kind of sense-memory possible associated with them (for example, one might recall them by appearance, by voice, by ways of motion, by scent or by feeling about being near them). They can also be connected with particular places in geographic knowledge or with particular events.

Character knowledge develops understanding and competence through recollection, discussing, observing, and imagining alternative possibilities. In this way minds make narratives that become active often talking repositories of knowledge.

Event knowledge associates regions of time in which things happen with knowledge or learning. Most people do not have a chronological, single POV sequence of real time unfolding action like unto a video. It's more likely to be fuzzy overlap of multiple related events that are mentally associated or seen as one.

Event memory is very good for gathering recurrent actions and blending them into an active region of possibilities. Recollection of events can reveal the why (understanding) and how (competence) of the event.

Going to the same place many times and doing similar things each time will usually blend into a single memory of what could happen there. This blending of event memory into possibility is what is used to create competence and understanding out of the awareness of recurrence.

Doing the same kind of exercise over and over can generate this kind of event memory, which in turn creates the competence to be able to do this kind of exercise again and the understanding to expand beyond the exercise.

Event memory can also contain traumatic memory, in which harmful and painful events are recalled with a combination of detail and augmented sharpness.

Traumatic memory can be hard to deal with and different practices work for different minds to clear it away.

Most minds have some combination of these three kinds of memory with varying degrees of emphasis. One mind might, for example, be very clear about places but nebulous about events and characters.

Some introspection will likely reveal which way you remember. And there is always the possibility that your recollection is not focused on any of these three.

Understanding Develops Knowledge and Competence

Understanding is an awareness of flow and possibility. In terms of praxis it connects cause and effect, dice rolls and outcomes, and the ways that processes broaden and narrow depending on circumstances.

Minds vary a great deal in how they are aware of understanding.

Understanding's simplest form is a back-of-mind-guiding-front-of-mind sense of what to do.

Praxes that have this kind of understanding feel natural and obvious and make the generation of competence relatively easy by front of mind trying to do what back of mind is showing to be done.

Such understanding generates knowledge by connecting things that fit together, building memory of how things should work.

More complex awareness can embody understanding into geographic, character, or event knowledge.

Sometimes understanding connects with particular senses, so that it has feeling or taste or smell or internal awareness.

Understanding that is being developed can create a sense that something should be tried or avoided. It can also manifest in uncomfortable ways like hypersensitivity, difficulty with balance, or normally easy tasks becoming difficult.

There is a related sense of understanding being under construction and temporarily taking away the praxis one is working on. This can depending on the mind lead to things like writer's block or bouts of exhaustion or depression.

Most people's understandings are pretty bad about time. Awareness can pop up at odd hours (This whole section popped into my mind at one in the morning, which is no fun, especially if one has insomnia).

If knowledge in geographic memory is like ground, understanding is more like water, flowing over and changing the knowledge, sometimes like rivers, sometimes like rain, sometimes like floods.

Understanding can flow over new spaces, turning otherwise neglected fields of possibility into materials and techniques for its arts.

Changes in understanding can reshape the whole interior world. These changes can feel sudden to the front of mind, but they're usually the result of a lot going on in the back of mind.

Part of the reason for the stories of eccentric inventors and artists doing weird things when working is that praxes of understanding have unpredictable needs. What understanding needs to work on and when it needs to do that work is very hard to regulate, though it's highly beneficial if one is doing a lot of making and remaking.

Minds have different reactions to the development of new understanding. It's not uncommon for the front of mind to temporarily reject new understanding (often accompanied by a thought like "That can't be right") or try to fit new understanding into a generally-accepted framework.

It's not a good idea to force understanding to conform. Understanding is personal. It will need to create new knowledge and suggest paths of competence that work for this particular mind. It will need to play and flow through its new configurations for some time before what has changed will become clear.

Once it is clear, the mind can discover that it knows new things or old things in new ways, or is competent at what it seemed incapable of before.

Competence Develops Understanding and Knowledge

Competence grows knowledge and understanding by increase of care, expanding what it tries to work on, expanding what it tries to work with, and expanding what it tries to work for.

Increase of care is usually a matter of slowing down action to get a better sense of what one is actually doing and what can be done to improve that. One should also consider the safety of one's materials, techniques and tools. Competence can guide the increase of knowledge and understanding in order to find out things like toxicity or risks of breakage.

Increase of care humanizes competence by taking account of risk and considering how to reduce it. Sometimes this motivates the creation of new praxis that will not have the unnecessary dangers of old praxis.

Mind expanding the space of possibilities that it tries to work on is the adaptation of competence to alternative materials and tools. This is one of the most common ways arts grow.

Sometimes new materials become available, sometimes old ones are no longer available. In programming, materials can transform overnight and need to be relearned. A language might be updated with a radical change in grammar that old software will need to be updated to accommodate.

New tools will pop up and arts will test them to see how they work. In programming this too can be imposed by updates. Changes in development environments can make it necessary to adapt one's praxes.

Competence can expand what it tries to work with by attempting to blend different praxes together. This can involve swapping one method out for another, such as using an electric mixer in place of a hand whisk or a recursive invocation in place of a while loop. Competence can also grow by embedding a technique into a larger scale technique, as happens with a great many cooking methods such as whipping frosting then frosting a cake, or doing two otherwise disconnected praxes on the same project, like glazing pottery.

This is one of the most broadening forms of experimentation and accounts for much increase of knowledge and understanding in the field of any art.

Lastly there is competence expanding what it tries to work for.

This is an attempt to grow what can be made with the art itself.

This expansion of competence often begins from the artist's eye seeing something and wondering if the art can depict it, and if so how, or seeing how something has already been depicted and expressing dissatisfaction with the ways it has been done.

Depiction here is a broad term. It can mean making clothes that show motions that had previously been suppressed. It can mean trying to use computer animation to show fur and hair that look more realistic without killing all available processing power and time in rendering. It can mean trying to replicate the effects of porcelain if one has no access to kaolin clay (Many European potters and governments tried without success to do this, although they did invent many new varieties of pottery).

All three of knowledge, understanding, and competence have their own ways to push individual artists to expand their arts. Letting the push of one's personal knowledge, understanding and/or competence happen and following it is an excellent way to grow as an artist and help the art itself grow.

Metapractices from Which Praxis Can Be Instantiated

Arts have large-scale metapractices that individual artists can instantiate to their own uses. Sometimes these metapractices are the names of the arts themselves – painting, sewing, embroidery. Sometimes they're smaller than the whole art, but still very big, such as wheel thrown pottery or database access and management.

Metapractices are generally taught with a toolkit of standard praxes that are declared to be best practices. As with all such claims, the artist will need to determine if the praxis works for them and if so what can be done to adjust it to their ways and needs.

Metapractices are often elaborated in tasks and teachings so that they can seem to dominate their aspects of the art as though they must be learned in order to learn the art. This benefits those whose ways of thinking can make use of the metapractice and can alienate those who don't.

This can create a problematic feedback loop within the art so that those who do not do the metapractice may feel that they should give up on the art, while those who do practice the metapractice will evolve and sophisticate the metapractice to do more and more of the art.

This can further escalate as the practitioners of the metapractice inevitably develop a localized art language purely for the metapractice. The development of SQL in the case of database management is an extreme version of this process.

This can lead to a lopsided version of the art wherein large regions of possibility are neglected because of overfocus on the things the metapractice does well and the conditions under which it does them well. At its most extreme this creates the idea that the one true way of the art is this metapractice.

The consequences of this can be far-reaching and potentially disastrous. Regions of the real world develop metapractices of farming which really only work in environments like the ones where they were developed. Transporting those practices to other environments with the blithe and ignorant assumption that every place can be farmed the way farming was done back home has shown itself to be very harmful to land and people.

Diversity of metapractices allows for arts to evolve more omnidirectionally and makes it easier for a diversity of people to take them up and continue to evolve them.

This is important because the conditions under which the practices of art evolve depend on the evolution of human need, usage and interests. Eventually the circumstances under which any art is practiced will evolve beyond the ability of any single metapractice to encompass.

These changes have happened many times in the history of art through the discovery of new materials, the loss of old ones, relocation of where the art is being done, the arising of new techniques and tools, and so on.

Arts in these circumstances can undergo catastrophic changes (at least in the mathematical sense of catastrophe theory, which models abrupt changes in process) wherein whole new metapractices need to be invented to deal with new conditions.

Arts with enough diversity experience less catastrophe and more shift of emphasis from one evolving metapractice to another. Arts that are more limited can generate conflicts between the old metapractice and new ones that are arising in response to new evolutionary pressures.

Ways as Clusters of Practices around the Artist

As an artist acquires and evolves their praxes by using them together, the artist will develop a way of doing the art. This is a feedback loop between the artist as a unique human being and the artist as practitioner of this art. The artist's personal way expresses itself in the art as what is called style, which is to say the surface of the artist's way emerging through their competence.

Deeper down, in the back of the artist's mind, understanding and knowledge from the different praxes are knitting together and creating the artist's evolving awareness and forms of action.

Their personal way can sneak up on an artist because most artists think they're doing things the way they do things. They don't notice how that way is changing unless they go back and look at their earlier work and wonder, "Why did I do that? That makes no sense."

This evolution that feels like standing still is a sign of the art and the artist getting closer together.

Done properly, this evolution can absorb ongoing alterations as well as catastrophic changes in how the artist perceives and makes the art, as well as which praxes they do, which they discard, and which they develop.

Play. Meant to Do That

Early on in learning any art, it is regrettably common to lose a lot of the enjoyment that might have propelled one into the art in the first place. Many learned practices and metapractices can be wearying. The work of adaptation can be difficult and the creation of new practices bewildering.

Good learning environments can provide opportunity for play, but in some environments and for some artists the commitment to learning can reduce the ability to enjoy making the art until the artist can develop enough of a personal way. Once an artist has the beginning of how they will do what they do, play can reemerge, energize the psyche, and enliven the artist's eye.

The artist's awareness and perspective on the world can change, sometimes abruptly, sometimes sneakily over time so that the art and its augmentation of reality become amusing. Experiments become playing with possibilities. The work can become both more carefully done and more enjoyable.

This does not always happen in ways that are obviously playful. Some artists get intensely focused on what they are doing and see the work as a struggle. But still their minds are rolling possibilities around and creating effects that they did not know were possible from what they knew they could do.

The rediscovery of play, whether the artist is aware of it or not, leads to a kind of skillful serendipity wherein the artist can generate unexpected effects by competently playing in regions of possibility and lucking into surprisingly effective compositions.

When asked about it the artist can say, "Uh, yeah, I meant to do that."

And in a sense, they did. They were playing around and something worked. But their play was dependent on what they could do. The result may have been random, but it was random chance in a space of possibility wherein nearly every outcome near their way of making the art would be good.

Having done this, the artist can examine the way they got the effect and incorporate it into praxis and so bring it back with care or more play in later works.

Following a river to new territory, one may find life one has never seen before.

Personal Exercises

1. Examine your learning cycles. See the rhythms and circumstances that help and harm you in learning particular praxes.

2. Examine any certainties you have about how to do things. Which ones developed from experience and which ones do you hold on to regardless of experience?

3. Consider what learning experiences you had that were illuminating to you and which ones you enjoyed.

4. What, if any, orderings of the competence-knowledge-understanding triad have you used and which have worked well for you?

5. Look at any specialty of any art you practice. See if you can find the metapractices of that subspecialty. Listen to the language used and try to gather in how it guides the mind toward using the premade praxes of the metapractice.

Testable Exercises

1. Take one praxis you do well. Describe your knowledge and capabilities in it. Then describe ways in which your growing understanding has changed your perception of the subject matter of the praxis.

2. Take a coding practice that works for you. Write down how to do it in a way that would help someone determine if it would work for them. Include what knowledge you rely on, what's easy for you that you rely on, and what's difficult for you that the praxis overcomes. If done in a class environment or study group, these practices should be shared.

3. After doing personal exercise 3. For at least three learning experiences describe whether you enjoyed the process or didn't and what if any aspects of it illuminated what you needed to learn. See if you can articulate what this says about how you learn relative to what you were given to learn from.

4. After doing personal exercise 4. If any such orderings have worked for you, write out formally how you have employed them.

5. After doing personal exercise 5. Write out at least one metapractice that doesn't work well for you and what trouble it generates. If you're okay with all the metapractices you use, write out how you make them work for you.

14

Artworks and Human Needs

What Humans Need

Our goal in this chapter is to examine artworks in general and programs in specific relative to human need. Developing an understanding of this makes it easier to figure out whether an artwork one is working on is doing what it is needed for, and if not how to improve its ability to fulfill human need.

From my perspective, the needs of humanity can be approximately classified as:

Feed the hungry

Clothe the naked

Shelter the homeless

Sustain the living

Companion the lonely

Teach the ignorant

Convey the weary

Each individual person has their own unique variants of each of these needs. Collectively, individuals and groups of people develop their own shared and sharing variants on providing for these needs.

At all times there will be needs that are not being fulfilled and needs that have not yet been identified as needs. The specifics of needs evolve based on how people are living their lives and what they have access to.

Humanity creating and audiencing art can be seen as an evolving process of need and fulfillment. This process exists on all human scales from the individual through family and friend groups through subcultures, cultures, societies, nations and humanity overall.

These scales entangle so that the needs of each individual are not lost in the needs of humanity.

Needs are entangled, not amassed. The needs of humanity as it now exists are made up of individual needs entangled into group needs into social needs and so on.

The needs of a unique individual do not disappear into the needs of families and social groups and societies and humanity overall.

In the other direction, the needs of the individual will always be shaped by their experiences of their families, social groups, societies and humanity as they take part in these structures during their lifetime.

DOI: 10.1201/9781003459866-16

Everything humans do is art. Every human need is a need for art.

Art also has needs that come before the art, the needs for tools, materials and inspiration.

Art as Tools and Materials

The need for art is also a need for the tools and materials that are used to make art. Those tools and materials are themselves also art because they are made by or gathered by humans for the use of humans.

One way to look at tools from the perspective of need is as a means to take one motion and from it produce the effects of another motion. Raise and lower a clutched hand, drive in a nail with the hammer clutched in the hand. Press down with a foot, a car accelerates. Click on a filename, cause an image to appear. Do nothing, a robot vacuums the floor.

The transformation of motion varies in how much of the movement is visible structure and how much is hidden infrastructure.

The hammer is pretty much structural. It is a direct usage of the momentum and angular momentum imparted to the hammer to drive the nail.

The car acceleration has some structural elements. The harder you press down the higher the acceleration. But your foot is not actually involved in the motive force, which comes from the infrastructural engine.

The file loading has a little structure just the selection or other input of filename, but the actual retrieval and display is infrastructure.

The robot is pure infrastructure. It can do its job with no human action or awareness.

When making tools for human needs, how much visible connection should there be between the human action and the tool's action?

There is no single answer to this. A tool that is mostly infrastructure does not require much human attention. This can be good if the tool should be usable without the human taking their mind off of whatever else they are thinking about. It can also be bad because lack of attention and carelessness tend to skip together carelessly off of unnoticed cliffs.

Tools can be made that both reduce the human work needed and create more awareness of what they are doing. This is especially true of software. As discussed before, we have the ability to show what is happening when a user takes an action without requiring the user to be physically hammering. The tool can teach the action of the tool while doing the action of the tool without the touch of human hands.

It is tempting for programmers to automate all tasks and hide all operations. Our field, after all, arose in the development of automation.

One of the precursors to programming was the Jacquard loom, which wove fabric into patterns created using cards with holes punched in them. The Jacquard loom shifted some of the emphasis in art of weaving away from the direct manipulation of threads and toward the creation of the patterns.

This reduced the hand work, but also reduced the ability of a weaver to directly affect what they were making. It sped up the manufacture, but took away the ability for the weaver to change the dictum as the res evolved in their mind.

What to automate and what to make easier or more controllable by an artist is one of the considerations of any act of tool making. There is no single answer because there is no single way tools work best. Each person has their own ways of using particular tools and those ways vary in which actions they wish to have their hands in and which they would like to not bother about.

When programming, it can be a good idea to present a scale of three to five options of how much the user's hand is inside a particular tool.

The most hand-inside setting is the most structural. At this level of involvement, the user can affect things more or less continuously.

At the middle setting the user can fiddle with settings but the system takes care of the work.

The most automated setting either just happens with no user involvement or works on a single control activation.

These different settings don't have to cover all aspects of using the tool. They can vary at different stages. For example, painting programs can have pre-made brushes and also let users make custom brushes. They can have painting tools which employ these brushes controlled by the user's hand and fill tools which automatically use the brushes in a defined area of the screen.

This gets back to the principle of play and learning.

Let your users play with the aspects they'd like to and learn how they work. Try to avoid forcing them to play with what you think should be played with and making it impossible to play with what you think should be automated.

Give them the space to develop their own praxis with the tools you've made so they can make their own art and enjoyment with it and satisfy their own personal needs.

Old and New Tools

There are four kinds of tools to make: Tools in common use that people need more of, tools that are improvements on tools in common use, tools that do known tasks in new ways, and tools that do tasks as yet unthought-of.

Making old tools again is just fine. It blends with tools that are improvements. People always need more of what they use a lot. And there are always small-scale improvements that can be made.

There are companies whose primary interest is in developing better hand grips for kitchen tools that have been in use for centuries, if not millennia. There are others who focus on making the tools themselves look and feel more beautiful for a given sense of aesthetics.

Each maker of old tools has an awareness of how the tools are used and should learn the history of the tools they are making. They may find that an older method can be brought back because its advantages have been neglected, or they can blend the old with current practice.

Improvements in old tools can be started by examining the assumptions made in the tools as currently employed. What does a user need to be able to do to make the tool work? What users are excluded because of this? What may be unnecessarily difficult for some users or unnecessarily confining for others?

Sometimes these questions can be answered by creating variations of tools that account for differences between people that are often ignored because of normalization. Left-handed tools are a boon to those of us who aren't right-handed. Wide and narrow sizes of shoes and clothes made for various body types benefit those who do not fit clothing designers' assumptions.

Some needs require the invention of new ways for old tasks. The development of speech and haptic controls for computers and the adaptation of interfaces to handle these controls have made computers more accessible to people for whom hand controls and visual awareness were difficult or impossible.

Completely new tools most often arise when someone sees human need that is not being filled. They see a disconnection, an empty middle space and build a method through the center using their personal awareness.

If you see such a disconnect, you can consider how a connection could be made and what would be needed to make that connection. If you can see a way, then you might consider it worth your time to manifest that way for the benefit of yourself and others.

Don't be afraid to invent, and don't assume that what is obvious to you is obvious to anyone else.

Making Materials

If tools do actions, materials are what they act on. Tools transform motion. Materials are what they move.

Materials have a different kind of beauty and utility from the beauty and utility of tools. The awareness of materials is an awareness of what one can do with them and what needs they might be able to fulfill.

Materials as a concept arise from the too-much-information of reality. Things in reality build up in complexity through interaction and entanglement and can then be taken apart in completely different ways.

Flax plants evolved to do what many plants do, grow in soil, water, and sunlight, and produce more plants. Humans discovered that fibers from the flax plant could be separated, spun, and woven to make linen, which can be used for clothes, painting canvas and sails. They also found that oils from the plant's seeds could be extracted and made useful for painting or as a nutrient.

Reality is like that. Too much information supplies us with an abundance of inobvious uses. Sand can be made into glass. Stones can be carved into tools or sculptures or gems or ground down for pigments or have metal ores extracted from them or be dressed and assembled to make buildings.

The thing that makes something a material is its ability to be refined and assembled in various ways by different tools and techniques for different arts.

Each stage of refinement and assembly produces materials that a person skilled in working with that material will be able to appraise and apply for different uses. A person who sews will learn about different fabrics and threads and be able to employ their qualities of texture and strength and flow and color for different garments. A jewelry maker will learn about things like metals, gems, shells and other hard, shapable materials that can be crafted and fit together. Cooks learn about raw ingredients and ways to transform them to cooked or pickled or just cut up and plated together in different ways.

Established arts have classes of materials that people can instantiate in whole new ways. The development of plastics by organic chemists created a new medium for painting – acrylics – as well as multiple new materials for sculpture and furniture and new fibers for threads and fabrics. The awareness of paint medium, of moldable sculpting materials, of fibrousness and fabricishness made it possible to see what could be done with the possibilities of synthetic polymers.

In programming we deal with three different categories of materials.

The first is *actual physical materials*. We can program automated processes that transform one material into another. Jacquard looms, programmable ovens, 3-D printers and robotic metal refineries all do this. Here we are programming tools to facilitate work with materials. A programmer working on this kind of code will need to understand the materials and the machinery doing the work.

The second category is the creation of *virtual materials*. The primary purpose here is to make it possible to do virtual experiments to see what would in theory happen if one did an actual physical experiment. These are used a great deal in materials science, various branches of chemistry, engineering, and various other branches of science.

Virtual materials are much safer and faster and cheaper to work with than physical ones. They can also be used in modeling things that are inconvenient to try to create in laboratories, such as supermassive black holes with star-eating accretion disks.

Here the programmer will need to understand the underlying science so that the scientists and engineers who will be using the program will be able to rely on the results of the virtual tests.

Everybody involved in this process will need to be aware that no matter how well a virtual material models reality it is still data modeling information. There is always more being omitted from the model than one thinks.

The third kind of material is the material we use in programming, namely *languages*.

Inventing new computer languages is a rare task, but an important one. If someone sees an empty space in what languages can do or in a possible approach to programming, they can create one.

This is a complex task and one that requires not just the theoretical creation of the language but also the writing of assemblers, compilers, or interpreters for whichever hardware they wish to use the language on.

The material of languages can be developed without inventing new ones. The creation of special-purpose libraries and APIs can increase the malleability of a particular computer language so it can be more easily used for a space of purposes.

APIs are most easily shaped by seeing what someone trying to do a particular kind of work might need to do and coding upward from the base language to that need. In this situation we start with unrefined language and combine and shape it to make metaphoric fabrics and gems and paints that are easier for other programmers to work with.

To make API programming a matter of materials, not tools, there must be enough flexibility in the data structures, functions and objects in the library for a programmer to adapt your creations to their work. If you restrict possibilities too much then you aren't creating materials, you're making a fill-in-the-blanks or color-by-numbers thing.

It's important to be aware of and respect the art of the artists you are making materials for. Do not code as if there is one and only one right way to use what you are making. Leave space for what you do not anticipate and cannot easily imagine being done with your work.

There is a certain delight in seeing someone make use of what you have made in a way that you would never have imagined.

Art as Inspiration

The ability to be inspired is an aspect of human learning. It amounts to taking something into the mind and from it gaining new insight into some aspects of mind and humanity.

Inspiration has been analogized to breathing in, to seeing light, to a flash of lightning, to eyes and ears opening and many other things in the history of poetry.

Inspiration can be the sparking of understanding through an interaction with knowledge.

It can also be understanding sparking knowledge by connecting or disconnecting ideas within the receiving mind.

Inspiration cannot on its own generate competence but it can guide the mind toward praxis that would help it cultivate competence.

Anything can be an inspiration.

Anything can inspire any awareness because inspiration is not the input.

Inspiration is the mind (usually the back of mind) acting upon, processing, and making new thinking from what comes into it.

Each unique mind reacts in unique ways to the diversity of information that comes into it.

Anything can inspire anything. But one can use art to bias the likelihoods of what will inspire and what regions the inspiration will arise in.

Art cannot determine how or what the art will inspire, but it can craft pathways and associations that can bundle sources of association together to make it easier for minds to be inspired by multiple elements of an entangled bundle of composed and contrasting information.

This bundling works by building knowledge structures that memory can associate together, often taking the form of place, character, and/or event.

Proximity and implication in art can more easily spark inspiration than distance and dissociation. Basically, a mind experiencing structured association is more likely to be inspired in some way by that structured association then if it has to go searching for individual pieces and put them together by hand.

Be aware that rage quitting can be a result of inspiration. It is quite possible that one's art will inspire disagreement, frustration and fury. It can also inspire a sense that, yeah that might work for someone but not for me.

Inspiration does not generate agreement and conformity; it inspires awareness and some form of understanding. That understanding may be avoidance, rejection or deconstruction.

The principle in making art for inspiration is to craft what the audience will react to, but not to rely on them having a specific reaction. If you make a three-step process and step two will only happen if the audience is overjoyed at step one then you will lose a portion of your audience before they reach step three.

The late cartoonist (among other things) Rube Goldberg was famous for doing comics depicting absurd inventions where one action would lead to another in a silly fashion. These inventions would often have people or animals in them reacting in stereotypical fashions. The inventions would not work if the characters depicted reacted differently.

Rube Goldberg inventions can be fun to see laid out in cartoons, but these are very hard to do in real life. There are contests for engineers to try to make some such devices that actually work. None of those real-world implementations rely on people having a defined reaction to events in the device.

If you're crafting art to inspire, you need to leave enough space in the work for the mind of the audience. You're providing information and assistance in processing, but they are doing the processing.

The art most likely to inspire is art that can be seen in multiple ways which each person can make use of as they need to at the current moment.

To give inspiration (which means breathing in), you must leave room to breathe and time for the audience to catch their breath.

Art as Food, Clothing, Shelter, Medicine to Sustain

Food, clothing, shelter, and medicine are all real-world information objects that humans need entangled with their lives to keep living.

Programming can't directly produce any of them, but it can be used in working to help people make them and acquire them. We'll talk about that in the next section.

But we can also look at these concepts as metaphors and ask, "What can our art do that acts analogously to food, clothing, shelter, and medicine?"

Food has taste and nutrition. Food provides people what they enjoy and need on a daily basis. It can also be a vector for poison and malnutrition and a generator of bad habits.

If this sounds like a lot of internet browsing, it is. Current internet software makes it extremely easy to ingest a lot of not-good-for-you stuff while trying to gather what you need.

We are currently in a randomized gathering society in which we forage through spaces that contain things we can ingest which might contain food.

But our current search systems do not give much ability to tell beforehand whether we have found food we need which is to our taste or stuff that's just bad for us.

The current tools are evolving and we cannot say what things will be like short or long term.

The base problem is that while gatherer societies could learn and pass on information about what they might be gathering, the internet is made of disconnected users searching through disconnected data rather than entangled information. We will not be able to replicate the food-gathering methodologies humans developed over millions of years because the environments are not analogous.

At the moment "feed the hungry" is an open problem. If this book is being read after it's solved, consider this section a historical artifact.

Clothing serves three real-world functions. It is protection from the environment we are in, it gives us something to wear that is constant contact with us, and it is revelation to ourselves and others of some of the motions we are doing. Clothing creates a changeable presence for us.

The interfaces of our software give us some shelter from the environment and movements of what's going on inside the programs. They also create a texture, a feel for using the program. Section II, Chapter 12, covers these aspects of clothing.

Interfaces also reveal our own motions to ourselves through their effects and feedback.

Revelation of our motions to others is mostly relevant in programs where we interact with others. At the current time the primary clothing in purely social programs consists of what we say, what handles we give ourselves, reaction emojis, and what pictures we use as avatars. In online games avatars can include 3D animated models, which can themselves have wardrobes. In games the actions we take and how we carry them out are also clothing.

Beyond this deliberate clothing, many programs generate clothing we are wearing that is largely hidden from us as users.

Programs that record user actions are generating records of motions. Programs that generate user profiles based on user actions are clothing the naked in garments not of the user's choice.

Such programs react to the user based on the garments the program has made for them, not on how the user wishes to show themselves. The garments made this way can have very unpleasant textures and can alienate users because they don't actually fit anywhere near as well as the programmers think they do.

In a practical sense, users are invisible. Our presences online are purely clothing.

If that clothing honestly reveals the motions and actions we take online then it is doing its job.

If it instead reveals what the program has crafted out of presumptions rather than real interactions, we are not being seen. A clothed puppet is being shown and claimed to be us.

Shelter as a metaphor is the shaping of the environment one is in.

Human senses of environment have a number of factors that don't show up directly in software – heat, humidity, aroma, and so on. But lighting, sound, sense of openness, narrowness or clutter, ability to maneuver, and how much and what is moving around us exist in the experience of computer use. All of these show up in the interface and in the ability to arrange the interface.

Shelter also provides a sense of proximity to other people. How much does it feel like one is alone and how much sense is there that other people are nearby and interacting with the same things we are?

There is also the question of hardware in an environment. What physical tools do we need to be able to control our virtual shelters? Do we need sound fed directly into our ears or into the real world around us? Do we need to be holding and looking at a portable device, or a screen, or do we need a VR rig or special glasses?

If we physically move, what does this do to our virtual experience? How does the real-world environment interact with the virtual environment? Have we got a full AR situation where two worlds are happening around us, both of which we are perceiving and interacting with?

Every piece of software contributes to each user's virtual shelter/environment. This is a case where the art will do this no matter what. It is better to give thought to the matter and act deliberately for what you are trying to make.

Medicine and other means of sustaining life can be seen as the taking in of information structures and processes that we cannot generate on our own but that we rely on to continue living human lives.

In theory computers can provide knowledge and capability to sustain our understanding by supplying knowledge and capability that we individually lack.

Access to knowledge and tools that can calculate or output data that we need to make sense of things is one of the great benefits of computers.

But just as what looks like food can be poison, so can what looks like medicine. There are large spaces of lies and many attempts to force people to think or act certain ways. Searching for knowledge and wisdom it is easy to find ignorance and folly dressed up in the garments of certainty and command.

The problem of making it easy to find what is actually beneficial in terms of sustaining life is the same as the problem of gathering food on the internet. Solve one and you will likely solve the other.

Art in the Flow of Logistics

If we shift from metaphor to reality on food, clothing, shelter, medicine, materials and tools, we can see the complex logistical flows that help in the creation and distribution of the things people need in their lives.

There are a wide variety of tasks within these processes that software assists with, from the finding and mapping of sources to the regulation of traffic, ordering, packaging and sending, pay and tracking of packages. E-commerce has made radical changes in what people can acquire and how they go about acquiring those things.

Logistics is the awareness of and arrangement of the infrastructural. Learning about it helps people see where and how things are made and how they get from place to place. It also reveals what materials, tools and skills are needed to make and transport.

The better one's sense of the complex web of cultivating, gathering, and processing, and the more one sees who has to do what to make and transport, not to mention what conditions people work under to do various things, the easier it is to see the entanglement of human needs and real-world materials and tools.

It's important to focus on logistics, not just economics. Economic abstraction loses too many levels of necessary human awareness. Money cannot magically transform into food, clothing, or shelter. The question of what is needful to fulfill need does not have a single-value solution.

To find the actual problems and needs, it's vital to see where all the human involvement is in any flow of creation and distribution.

Parts of this complex web are visible from the perspective of each person doing a part of the process. If you examine and connect these perspectives, if you see each person's needs and each person's interactions with other people in the logistical flow, you'll be more able to see who might benefit from your work and how.

There are a number of praxes that can be developed in order to be able to create this awareness. You can start with any object you own and try to track it backwards to find how it got to you, where it was made and by whom in what conditions, where the materials they used came from. You can learn about the tools used in the making and where they came from.

You are not likely to be able to answer all of these questions, but you should be able to answer enough of them to get a sense of that piece of the web.

Do this with a few different objects and you may start to see overlap in how and where things are made, how they are transported and by whom.

Now look into what software is involved with the various tasks and who uses it for what purposes. This will give you another layer to the web with which you might be able to see some empty spaces where tools are needed and others where systems could use redundancy, or alternative sources might improve things.

There are larger-scale layers in the logistical awareness of the condition of the world and environment that affect many human flows of materials and art.

Weather modeling and prediction software and climate modeling are part of the short-term and long-term factors that cause the web of logistics to change. Political alliances and treaties can also affect who has what available.

No part of this web is stable. Weather is chaotic. Climate has catastrophe points. Politics and governments depend on shared human actions. The layered systems depend on continuous input, ongoing correction, and awareness that a great deal can change in short periods of time.

All things evolve. Understanding how that evolution happens and what its consequences can be is necessary. In order to have a clear sense of the current and possible future arrangements of and for humanity, we need to build for what happens to be and what can possibly emerge from what happens to be.

The web of logistics is one of the most augmented-reality parts of human life. All human lives and possibilities flow through it as it flows through them.

You may not end up writing any code for it directly but all the code you do write will be involved in it somewhere.

Art as Teaching

Teaching is the action of providing possible knowledge, sources of inspiration and guided exercises in order to help someone develop praxes.

Any piece of art can create knowledge by creating specific associations within it and by its placement in reality. Any piece of art can inspire. Any piece of art with narrative or flow can provide guided exercises because a mind following it can create associations through that flow. Any piece of art can teach.

This is not the same thing as constructing art for the purpose of teaching.

As art evolves and humanity learns and experiments and as science advances, it becomes necessary to create and update the materials and tools which teachers use to teach.

New tools can cover advancements in knowledge, understanding and capability, or expand teaching methods to help those who had trouble learning using currently used methods.

It is not uncommon for someone who has learned something in a way that was hard for them or has seen others struggle with learning things that are easy for them to see a need for alternative teaching methods and tools.

Art made to teach deliberately selects sources of inspiration, which may be individual works of art or illustrations or places of possible interest or access to audience-selectable sources of information such as libraries or museums. Art made to teach can provide overlapping regions of information that can be used to create knowledge. It should have exercises that help students develop their own competencies.

Exercises should help students find and develop their own approaches and timing to what they are doing.

Some can be quite general, such as "Do three drawings a day" to help generate the habit of drawing.

Some can be focused, such as teaching a complex procedure by giving exercises for each step along the way.

There should not be any single exercise that must be done in a specific way to be able to progress. Exercises are stepping stones, not stumbling blocks.

Books and software that teach or aid in teaching need enough flexibility to let a variety of people learn from them.

Ultimately any such teaching source will only work for some people and will work uniquely for each of those people. No tool or method of teaching will be universal. Students and teachers need to be aware that they may need to go beyond what they rely upon in order to have this student learn this subject from this teacher.

Crafting artworks for teaching requires presenting essential ideas in diverse ways. Even if those ways do not work for every student, the diversity itself can illuminate unstated and unshown ways.

The style and humor of any source of teaching can strongly affect who gains anything from it and who doesn't.

There has been, going back at least as far as the Spartans and the Romans, a mythology that strict following of a single optimum sink-or-swim path will create a perfect cadre of hypercompetent graduates who can conquer the world (literally or figuratively).

This was a myth when the Spartans and Romans said that's what they did, and it's a myth when their fans 2000 years later try to emulate it.

Bluntly, sink or swim teaching does not work.

It produces a small set of people who all agree with each other about a field and a large set of people who give up on a field that they could have become part of had they been given other means of learning.

There has also been an opposite mythology that claims that students work best with no guidance whatsoever. This mythology misses the point that part of learning is gaining the benefits of learning from other people's mistakes, not just one's own mistakes.

In the actual learning space of methods which can work, there is room for individual humor and guidance from a wide variety of teachers, each with their own approaches, especially if those approaches can handle a wide variety of student approaches.

On the whole, software and books that have a sense of approachable personality to them can make it easier for more people to learn from them. The teaching artworks don't need to feel fully companionate (see Art as Companionship below), but many students benefit from teaching materials that don't feel distant and imperious.

Everything we've been saying about awareness of possibility and differences in user approaches goes double for teaching software.

If you are coding such software, give it a presence that the user can be aware of, and give it enough illumination of praxes that users can interact with it. Don't just make a bundle of graded exercises. If the user/student can pick the teaching up and play with it or generate their own exercises, then they are more likely to stick with the teaching and learn from and through your software.

Art as Companionship

Conversation, shared play, enjoying art together, and discovering unknown commonalities are all arts humans use to become closer to each other.

Direct Companioning

Software has extended the reach of all of these arts. The ability to converse across the world, to play together on the same machine or in local networks or online, and the ability to share art worldwide have all made it easier for people to interact.

These same software tools have also made bullying easier and more far-reaching and have created new kinds of danger people expose each other to. There is a need right now to make anti-lonely software that can't be turned toward making its users lonelier.

The basic problem lies in the fact that companioning does not scale. Companioning is a need individual humans have for sharing space, time and possibility with other individuals. It is a person-to-person or small-groups-of-friends need.

Large open spaces for people to interact in can make it easier for people to find people they can interact with in a companionate form. But interaction spaces are easy to overwhelm with groups of people who are all loudly saying the same thing. The harmony of human interaction is disrupted by the noise of large-scale bullying.

Social networks turn the spigot of human interaction all the way up to flood levels.

The most common effective solution to prevent overwhelming harm is active moderation policies.

But moderation is a human activity and cannot be accurately done by software. Moderators often experience a great deal of personal distress as they try to filter harm and harmful people out of a system.

Moderation policies can be enforced to protect bullying and remove protection from the harmed.

It takes a lot of human work, compassion, and willingness to fix messes to keep companionate spaces operational.

This is a space of open problems that many programmers are working on at the moment. At least one generation of emerging coders are growing up with direct experience of this need. They may not know it, but many backs of minds are bubbling through possibilities. Also, many of us older programmers who watched the evolution of this situation from the beginning are working on new approaches.

Right now a great deal of information is being gathered and a great many ideas are being contemplated.

Unfortunately, a great many people are also being hurt while the problem is being worked on.

Indirect Companioning

Many people find companionship in spaces that are filled with art, such as gardens, parks, and museums. Others feel companionship in buildings with interesting architecture and views. Software can provide virtual tours and modelled and animated spaces that provide such. There is also a feeling of belonging that some people get from amusement parks and theme parks which can also exist in virtual forms.

Most such places, real and virtual, work by giving a sense of being able to wander and explore, to find one's own way through a safe place and therefore create a personal memory of the space. Virtual wandering lacks several of the senses of real life, but the environments are less likely to have wild animals, mosquitos, and dangerous terrain and weather.

Immersive art such as books, sequential art, videos, and games also provides a sense of companionship to some. Such artworks provide spaces but they also usually provide characters and events and therefore can appeal to a wider variety of memory types than virtual tours.

Music can also companion people. Song often gives feelings of closeness to the audience because music stirs feelings and song creates memories.

This can lead to a parasocial sense of connection to musicians and singers that is not human-to-human companionship. The audience is being companioned by the art, not the artist.

And yet the feeling of mental proximity and understanding that generates that illusory connection is an actual feeling in the mind of the audience. The lonely within them is removed, even if the human who created the art that removed it is not there and does not know it happened.

This is an important distinction. Loneliness is an experience that can be removed in many different ways. Not all of them are symmetric relationships of mutual caring. Some are distant interactions where one person's art has helped another person feel closer to humanity even if they never meet. This can even happen if the artist made their work centuries before this person ever experienced it.

Companions can also be made from toys by people projecting their imaginations onto them. Children do this with stuffed animals, dolls, action figures and indeed many surprising objects, as humans can project humanity into anything. Older children and adults can garner characters from fiction or games, creating mental companions that dwell in the reader/player's memory.

Many computer roleplaying games actively encourage the development of companionate relationships with non-player-characters by creating dialogue trees that respond in different ways depending on what the user does in various situations.

Recent software has escalated the complexity of creating imaginary companions by using language learning tools.

Learning software is employed to generate response structures that learn relative to what the user does and how they react to what the program says. This software makes it easier for many people to project onto and craft companionate interactions. This software is still very new and as with all new tools and toys the uses and pitfalls are being explored.

Companionate software has been employed in robots and this field is also very much in its infancy.

The important thing to be aware of in this field is that humans can make companions of anything and are capable of projecting humanity on to anything. But they cannot make something be another actual human.

This field of software relies on that audience capability more than on any artistic ability to accurately replicate humanity.

Art That Conveys the Weary

Humans become wearied by overexpenditure of energy and often by overinvolvement with work and sometimes by spending time with other people. Restoration of weariness can come through a variety of means.

Rest and sleep are the most obvious form this conveyance. Mental passage through fictional spaces, for example reading, watching videos or playing games, can also accomplish this, as can listening to the kind of music the person enjoys. Certain kinds of vigorous exercise such as dancing, swimming or running can remove certain aspects of weariness even if they do not remove actual bodily fatigue.

Software as an art is very well suited to this kind of unwearying.

Conveyance from the weary mind works best with regulated persistence wherein things change at controlled paces and according to knowable ways.

Properly-constructed software environments can create conveyances where people can easily spend hours immersed in unreal spaces and worlds implied by sound and animation and emerge refreshed to face either reality or different unreality.

Worlds like that do, however, require more arts than programming to make them live. Which is a segue to our next chapter.

Personal Exercises

1. Do testable exercise 1, then consider what the software needs of at least one person in the logistical web involved are. Find out, if possible, what software they used and consider what you might do to improve it.

2. If you find the listing of need types does not fit your thinking on human need, consider how you think about it and consider how your experiences with art fulfill or don't fulfill your needs or the needs of people you are close to.

Testable Exercises

1. Take one object you own, track down where it came from, how it was made, where the parts and materials came from and which people in which places did the work to gather or produce those parts and materials.

2. If any of the open problems outlined in this chapter about software as need fulfillment interest you, describe how you see the problem and what aspects of it you feel it might be possible to tackle. This exercise can also be used for class discussions.

15

Combining Coding with Other Arts

Sometimes the data we need for our software is the work of other arts. Sometimes our code is largely a showcase for the work of other arts. And sometimes, as with cartooning and theater, a single artwork is inherently the work of multiple artists practicing different arts simultaneously.

Each art has what it does best. Each art has its own variation of the artist's eye which reveals the world relative to what that art does best.

When different arts are brought to bear upon the same subject, they can compose and contrast their eyes and their workings.

Done properly, the contributions of the different arts can entangle to make works that are richer in information and implication and more enjoyable for their audiences.

How do we bring coding together with other arts to make such works?

If You Are Also the Other Artist

There are advantages and disadvantages to working with yourself.

On the one hand, you understand the res you're trying to create. The back of your mind can perceive the work from the perspectives of different arts and make manifest what is needed from each art you practice. This means that most of the time you don't need to explain your needs to yourself and you have a clear sense of what's needed for the job and how long the various tasks will take.

On the other hand, you may need to develop some hybrid praxes because the better artists in the back of your head may not automatically synthesize understanding. Your knowledges of your two arts may conflict in ways you didn't see beforehand.

Suppose your arts are coding and painting and you need images for your program.

As a painter you may be used to creating large images that you expect your audience to be able to linger over, examine, and gather from over a long period of time.

But your program might need 20 images, each 400px by 400px, which your users will make use of as icons to help them be more focused on the actions they are taking in different parts of the program.

In short, you need to create memoria that are not supposed to catch and hold the user's attention.

Do you change your painting praxis to accommodate this need?

Or perhaps you should change how the program will use the images so that the user will be able to linger and consider and gather more understanding from the icons.

Do you make a blended praxis that creates paintings that do more work in the program than just act as icons?

DOI: 10.1201/9781003459866-17

Suppose you are a writer used to creating novels and you are creating a game that will have little bits of lore show up in fragments as clues to an overarching background.

How do you create the broken-up pieces of writing so that they thread well through the game plot?

Suppose you are a composer and are used to making complete pieces of music that will be played from start to finish, but what you need is background music for a game wherein themes will show up to warn the player of imminent danger or intense NPC interaction.

As the second artist you will need to adjust how you work to be able to make what is needed.

This can involve new praxis creation and will take some effort and care. You may need to do some deconstruction of the assumptions in the praxes you already have and entangle in new assumptions from the other arts involved.

Once you've rebuilt your praxes, you should be able to reap the benefits of being both artists involved in the making.

If you end up doing a lot of this hybrid work, you may need to rebuild some of your coding praxes so that there is more room for your other art to influence what programs you make and how they work.

If Someone Else Is the Other Artist

Respect your fellow artists.

Do not be a bad client. Before you talk to another artist about what your program will need from them, try to figure out your needs in a way that you can explain to the person you are working with. Do not be demanding and controlling. You need to give their backs of minds room to work.

Your needs should be the inspiration for their work. Give them a clear understanding of what your needs are, while leaving space to be pleasantly surprised by what they do.

Depending on the way the other artist's mind works, they may have questions about your needs which you may not have considered. Talk things through. Give them what they need to work with, and this includes time and effort.

There is a common error people make, which is to not know how long other people's work takes. If you don't know the art and don't have much experience with artists who practice that art, you may, as many people do, think that what you need will be quick work.

Take the time to count how many individual artworks you need.

Count them up and consider that each is an individual work of art.

Develop a sense of what you are asking for and how much work is involved.

The easiest way to do this is to ask.

Depending on your work flow and design, you may be able to have much of the art made in parallel with the programming.

As the res and the dictum of your program evolve there will likely be artworks you didn't realize would be needed or that the works you do have may need modification.

Leave room for the possibility of change. Understand that each change is something the artist you are working with will have to implement, which will involve different amounts of back- and front-of-mind work on their part.

Be aware of what changes you really need. Don't change your mind whimsically.

Every art has things that are easy to adjust, things that are difficult to adjust, and things where the artist will need to start over. Learn the difference between these for every art you are interacting with.

Understand that two arts that might resemble each other can be very different on the inside.

A scene made using 3D modeling can be looked at from multiple directions and can have its lighting changed with provided tools.

2D paintings and drawings have fixed points of view and individually generated lighting effects. They can't be modified in any simple fashion. For 2D art a simple-sounding change of point of view requires starting from the ground up. A change of lighting can involve drawing or painting over large swaths of the work.

On the other hand, 2D drawings and paintings can in general have new things added to them more easily than 3D because it's easier to draw a bird at a distance than it is to create a bird model and position it just so in a 3D space.

What's easy for one art can be very hard for another.

Learn what the other art does well and what it has trouble with, so that if you need changes you can request changes that are simple for that art.

As we will discuss in Section III, Chapter 23, art is copyright to the artist unless very specifically contracted otherwise in specific legal terms. Copyright may be sold or licensed. It does eventually expire, after a very long while – 95 years from first publication or 70 years after the death of the original creator, in the United States at the time of writing.

To use art that you did not make you need to have the legal right to incorporate it into your program. It's a good idea to have contracts with each artist you are working with.

Various Arts that Are Used in Software

2D Images

Pictures are so incorporated in our current computer experience that it takes me some work to drag my mind back to the time when everything we were doing was text and we had to make ASCII images out of text characters to create anything close to a picture (Figure 15.1).

Old Fogey alert: The first long program I wrote was a spaceship combat game where sectors of space were depicted as a square array of dots and ships were letters that took the place of dots. Missiles in motion were shown as *.

This was not shown on a screen. Each move was printed out on a continuous roll of paper.

I'll stop rambling now.

Skip ahead half a century and we use 2D images everywhere.

Incorporation of 2D art in current programs is facilitated through various interface elements. These can be idiosyncratic to language or operating system, but many of them just involve specifying what image file to put where in a window and what size to make it.

There are a number of common uses for 2D art in software, each with their own considerations.

```
                       .*C+CTF&OCI+;:..
                   .COMMIINMMMMMWMWBEHEGC,.
                .JBWMMJIHWMNMWMMNMNWWBEBKC;.
              .CWWWWGHYLEWMMMMMMNMMWMNWMBWHRL.
            .JMMMWYC/...VTSQRHBWMMMMMMMMWMWERZ\
            .AMXJIL;;---;;+;/JCOHMWWWMMMMWMWNKO;
           ,ABF*=:..........,,,;;YFEMMWORBMWNMHBPE&
          ;ZWGJ;,............,,,;.VTHBWMYOSHBWMBEBP,
          /JWBV;,.............,,,;+JVFHEBWMW8FZWMBEH:
          JIGMD:,..............,,;;+JLTPZMMMWRTPMWBED.
          J*WMC;:,...............:=IIJYITHMWMMBFZDWER.
          I»HMW;;:,,.............,,,,;;;=JTHWMMMWBCBWNH;
          L+KMMI;\;,:-,....,:/«=+»/;::;IJLTBWMMMWBELWEI
          I/RMWSYPFTTFYL..-TFTY**YZREH&CTOWMWMMMWBJLWI
          YKWMIIF.IMBLI..JG/:.WMYLKYLIIJTBMNMWMMWB&L7
          /ZNWMWI.TFYHKLI.,JSY*IJVTY/=*IIJYBWMMMWMMWML\
          JOZHNWM;,..,:»7;.JCYI;;::::;»+/IIJSEWMMMMMNBMWEA\
         .COZBWNM«,...:»I,.JVLI=;:::;;»=**JSEBWMNWMNBWHKI.
         .VTDKWBMA:,,,;/*..ICLI=;:,;:»+*JCSEBWMNMNWBENKOL,
          CTFRWWMBI.-/IJI.,JVCLI....:+*JOHEMMWNMNWMNGZFII.
         .VFAKBWMBZ*\;:.TLAKWNEL;.=+*JZGKEWWNMMMNWMBHEOIL.
         .«ZXEWBNMLII*.,.;ITTCCO&+:+*IYGENWMMNNMWBHHKAFLJ.
          ;OXHBWWMB&**JZALAAXSQEKT;:*IJDGWBWMMWBNHAEGF&6L,
          :IJCZHBMMS&**-,-/=+CLL+..+*IJSENEMWNMWEHGRK&FVJL.
          :IVTYEBNMWS&**;*OPFLII**/»IJYENMMNMWNMWBAERA&CLII
         /IDZPRENMMWAP&*«..:+IJL*IOZDBWNWMWMNMBEB8RDKHYHIJI
          ;JTOFSEWNMMWBKZL=+*JF&KRHEBEWNMMWNHWHKRAKSHKXKI»I
          :/JPZGHNNMMMMMWEADSZHKRHBWEWNMWNWNBEBRHKKXPATHABI/.
          :JLV&HBWBNWMWMMMMNWMNWMNNBEBHRXSTSDRKPHBABFSGR/.
          ;JVPREBWWNMMNMNWI;\YDRHBEBDSDAGIVBHWEWHRHEDAHI
          .\FVKBWMNWNMMWMNSL;;YCTFPF&FPOTY«YBBHNEKHZHTYH.
           IJOZDHBMWWNMNMWHI,\IJIILIJJIL/IISHNBEBXEPBWKBL.
          :ICHBMMNMWMNVI«/:;**I**I*+*+*I*IJYKARHJRIWHAWNL
         /CGHBHWMNWBHTYI«;;;;**+*+**+*I+++*IRXTEJHIHBXEWBL.
          ;JKEBEWBEAFI=JI«=«=«**=«*+=«*+*I*JITZPIKIJKEABBWERA,.
         /«CT6XRHKTI=«+;;+;;;;+*+**«*«;;;*IS;YFGISIVYHDEHEHRWKO\.
          .IZSKAOPFI;;;;:::;+::::+;;::::::;;IVIY.S*Z+LTVTFOTFZHEHZF.
          .TDKRLJTI::,:,,,:,,,:,,,:,:,;,,,;IVIJIJ.P/CZJOZPDSKTCGHRO.
          ..AY.J/:,.,.,.,.............,.,,,,.:Y.....FZFJTLIF;,,:;IJTFO.
          .Y;   ,  ..  .............  ....,,,.,VI...IY....            IF
                                        ;.;/,...                      .
```

FIGURE 15.1
An ASCII image.

Background images and loading screens are usually pictures, sometimes lush, sometimes simple, meant to fill an entire screen. Background images need to be behind the parts of the interface that the user is interacting with.

Background images can be interesting or plain, but whatever they are, they should not make it harder for the user to do what they're doing. Background images should not blend in with cursors or make text muddy and hard to read. Loading screens can be just about anything as long as they don't make your users want to look away or quit your program.

Icons are images created usually for controls to act as memoria. They should be easy to spot and make sense of. They can be derived from common symbols originating outside of computers. The use of the octagonal red stop sign icon as a warning is imported from driving. The rewind, pause, play and fast forward buttons used in most video player apps entered from the physical buttons used on VCRs.

Icons as memoria need to be distinctive.

They don't have to be simple. They can be fun or playful or dramatic. They can even be enjoyable to look over. But they can't require that the user stare at them to figure out what they mean.

Avatars are like icons, but instead of marking out controls, they are used to stand in for humans, other creatures, and characters. Avatars can be a lot more complex and subtle than icons because living things are more complex and nuanced than pushbuttons.

Avatars should not be very large.

Avatars are usually coupled with game actions or someone's comments in a discussion. They should make it clear who is doing or saying what, but they should not overshadow what is being done and said.

If your program will have a lot of avatars, you and the visual artist will need to consider how similar the avatars should be. Stylistic similarity of avatars can be used to help promote a unified aesthetic. Stylistic diversity is useful for letting people determine how they wish to be seen and to tell characters apart.

Maps and searchable spaces are large images that contain areas users might wish to examine or travel through.

Maps abstract from reality (or imaginary places) to give an overview of what is happening where. They act as a step back, making it easier to plan actions and integrate a larger awareness. Maps can have legends, which are icons or words in various places to make known what is there. Maps that are also searchable spaces will have a means of finding things on the map and will change their legends based on what is searched for.

Searchable spaces contain interface elements in particular places. Technically all windows are searchable spaces. A picture of a room in a hidden objects game is a searchable space. The picture has regions representing objects and tapping or clicking in such a region will usually highlight an object or cause something to happen to change the image in some fashion.

Such images generally need to be made in a style where objects are sharply distinct, so that the viewer can make out the possibilities.

Galleries are a specialized form of searchable space that acts as an environment for presenting pictures to the user.

Galleries can be employed for simple purposes such as showing off various art collections. They can also be used as a major component of games wherein places are depicted as 2D images with character icons or dialogue imposed on top of them.

The important thing about galleries is that they give the widest space of possibility to the visual artist. The program is shaping itself around the art, not the art around the program.

Writing

Writing, the putting down of words and symbols in a form that can be interpreted as meaningful, is entangled with coding at a very basic level.

We're going to skip all of that because they're two very different arts and what we're focusing on is how and where these arts come together.

There are two fundamental connections between these two arts: Writing about coding and writing that appears inside programs.

Actually, even those aren't clear cut separations. But let's go with it for now.

One form of *writing about coding* is writing that appears within code itself, namely commenting.

Commenting is explanation for what a piece of code does and how it goes about it. Commenting can help programmers understand each other's code and to some extent how that other person saw the code they were writing. It can also help some coders when they go back to fix, update, or scrap earlier code.

But commenting doesn't help everyone. For some programmers, comments can interfere with the ability to read the flow of code. For some of us reading code is like sight-reading music. It's harder to read music if each bar is followed by an explanation of what's happening within the bar.

Nevertheless, commenting is a commonly used praxis because it helps many people. Just be aware that it can also get in the way.

Comments should be clear about purpose and process, but not too distracting from the code itself. Remember that the audience for your comments will be other programmers who know the language you're writing in. Keep your comments focused on what the code is doing and what part it plays in the larger program.

Put your comments where they'll be most helpful. For example, a brief description of what an object is for can be placed at the top of the object declaration.

These days the most prevalent form of *writing about coding* is found in online discussions of problems coders are having getting languages and hardware to do what they need them to do.

This is a valuable communal activity that coders do for each other.

If you are in need of such assistance it's a good idea to state your problem clearly. "I'm trying to do X in language Y on system Z. I've tried A,B,C (code snippets), or the manual says do it this way (code snippet D), but I'm getting results F,G,H or error messages E1,E2,E3. Please help."

Most responses should be of the form "I've had some success trying I or J or K" or "Yeah. This is a known bug with this implementation. Here's a workaround."

Try to be concise and helpful. Don't second-guess the methods the person is using or insist that the person is doing it wrong (unless they actually have a syntactical or semantic error).

The largest-scale form of writing about software is books about languages and how to use programs.

These are of two types.

There are manuals, which are technical writing laying out the exact syntax and effect of each command or control within the language or program.

And there are how-to books that focus more on what can be done with the language or program under discussion.

It used to be the case that people were expected to read the manual before actually running a program.

This didn't work. People just dove in and started playing around and looked things up only when they had to.

So the psychology changed. Physical manuals are rarely supplied with programs these days. Users are expected to employ built-in help or to search online for how to do things.

What used to be sections of manuals are now written as separate online articles that are hyperlinked together and can be used as help articles, rather than as single books to be read from beginning to end.

This has the advantage that users have an easier time finding the particular pieces of information they need, but the disadvantage that they have less sense of the overall capabilities of the programs they are using.

It can be difficult for users to find the particular help article they need because it may not be searchable using the ways they think about it. Writing help articles therefore has the added difficulty of figuring out how your audience might reach the need for this help.

Writing within programs mostly consists of legends, lore, scripts and active stories.

Legends in this case means labels on controls and maps.

Most legends in software are flat technical labels without any association. There's little to no poetry in the labels given to controls and the legends that show up on things, and that's unfortunate. Poetry makes memory.

Controls need to be distinct, but they also need to be capable of association with each other. If your controls feel like they belong together, the user is more likely to perceive your program as a unified whole, capable of doing what they need.

This kind of poetry shows up in controls that are metaphors for real-world things. Drawing programs use poetry by calling controls "brushes" and "erasers" and "paint cans" and "palettes." "Cut" and "paste" as commands are poetic associations. You can help the user compose their awareness of what your program can do for them by bundling your controls by associations.

Lore is the usual term for chunks of text in games that convey backstory or explain quests or other activities that should be done. Lore can also be used in programs to explain controls and capabilities and why they exist. In both cases the lore makes the program and its contents more real to the user and helps them generate a personal awareness and attitude toward this aspect of the program.

Lore should consist of relatively short but illuminating pieces of writing. The amount of time they take to be read or listened to should amount to a pause in activity, not a stop. The user should be able to take a moment, examine the lore and return to what they are doing. The time taken should not be long enough for the user to have to ask the question, "What was I doing?"

Scripts are structured dialogue.

They may be responses the program gives to user questions or interactive trees within games or suggestions that pop up based on user actions.

Most scripts are prewritten. Some can be generated by language learning systems, but that's really a software-software interaction rather than a software-writing interaction.

Scripts can have a question-and-answer format to gather information from users in order to determine how to process their needs and requests. This can be useful in order to make it easier to generate the contents of a form (Tax preparation software usually uses such a Q&A format).

Scripts in games are written to fit the personalities of the characters speaking them. Writing these is an adaptation of standard dialogue writing, just with more possibilities. The writer needs to figure out how character A will react to the user doing action B and write accordingly.

Scripts between programs and users, such as filling out forms, are often written either in an impersonal style or a pseudo-friendly style which can be either comforting or off-putting depending on the user and their situation.

Tone and helpfulness matter a great deal in these situations. A person filling out a health insurance claim or a report of a car accident is likely to be in a different frame of mind than someone who is trying to book a family vacation. Scripts need to be written with possible user awareness and situations in mind and give a sense that the program was written with that awareness in mind.

Active stories are something of a cross between lore and stories. They are used in games to chart the path or paths that the user will be playing through and to involve the user in the story of the game. Active stories can show up as main paths and side quests. The general structure is a story with possible branchings and scripts embedded in it, as well as places where game actions need to be done to advance to a later part of the story.

This has a resemblance to Shakespearean playwrighting, wherein scripted speeches are broken up with phrases like "They fight" or "They play," or "They fight. The duke is slain."

The equivalent in a game of, "They fight. The duke is slain" would be the story sets up a fight, the game handles the fight, and the next part of the story happens only if and when the duke is slain.

Games vary in how connected the game and story actions are. There are a variety of puzzle games in which the plot stops for the user to solve a puzzle which may or may not be something happening in the world within the game. Until the puzzle is solved the story does not continue.

The relationship between the game play and the active stories of the game should be worked out between the writer and the programmer. Good interconnection and a sense of player agency in the process tend to make such games more satisfying.

Online Design

While HTML is a programming language, its origins were in the art of layout design. So, the people who studied and made use of it often came from that region of visual art rather than from a programming background.

HTML evolved layout into the skill of web design which evolved to include more programming as HTML, Javascript and other web-affecting languages and APIs evolved.

It is possible to look at online design as programming that goes from the outside, in that is it starts at interface and moves inward to code.

But this misses the difference in artist's eye. Online design focuses more on user perception than on code capability. It comes from the space of magazine and newspaper and advertising layout and walks into the space of programming. People from this area use many of the same materials and techniques as programming but with very different eye and purpose.

It can be tempting to think that a good interface coder has no need of someone with the skills and perspective of an online designer. But that's not really true. The differences in artist's eyes and artwork goals can either conflict or can feed each other so that the coder and the designer can mutually craft what the code is doing.

To make this work each must respect the needs, concerns and skills of the other. Unlike other art interactions there are no specific considerations here. The arts are too firmly entangled for separation and clarification. Two people are talking about the same thing using two different languages each with their own biases.

They can meet in a middle they are mutually creating if they pay attention to the fact that they are talking about the same things. The important thing to be aware of is that each person's viewpoint is needful. Neither can really do this job alone, because ultimately both directions of view are need. The truth of one is the beauty of the other.

Sound Effects, Music, and Singing

Early on computers could only activate connected noisemaking devices such as the physical bells on teletypes. Technology advanced so that computers could synthesize very simple sounds, then advanced more so that it became possible to create noises of various types and then record, play, and generate real music using Fourier series.

Sound effects are popular. They are employed in games and other programs for alerts, evocations of real-world noises, and comedic effects. The art of creating sounds that sound like things is called Foley. It is a skill founded in the ability of the mind to recognize a sound, and by association imagine a thing that makes that sound.

Foley evolved well before computers and sound synthesizing technology. The skill is still in use because a competent Foley artist can figure out what noises to make to evoke audience imagination regardless of whether the sounds come from banging coconut shells together or a synthesizer.

Music brings feelings to life. The music used in software can either be recorded from live musicians or created using virtual instruments playing according to coded instructions.

The latter is in theory more flexible because the instructions can be changed on the fly to fit different situations. However, programmed music lacks the ability of live human artists to play according to their personal and shared interpretation of the music.

Both coded and recorded music have uses in programming, primarily in games, where leitmotifs for characters, places, and events and full-scale orchestral scores can be used to give emotion and richness to action and environment.

Creation of such requires a composer to score the music, and, if one wishes the benefits of human musicians, one or more human musicians.

Be aware that the composition and the recording are legally distinct artworks with distinct copyrights. Having the rights to one is not the same as having rights to the other.

Discussions of what music one needs for a given situation can be difficult because evoked feelings can be hard to articulate. The language we have for emotions is far simpler than the complex interplays of the human mind in motion and the music that calls it into being.

It's also important to remember that the music that most affects you on a certain level may not do the same thing for other people. People often imprint on music that they heard when teenagers and that can shape lifelong attitudes. One generation's deeply affecting compositions is another generation's stop-making-that-racket.

For this reason, the kind of music you are using in your software can have a strong selection effect on your audience.

Be aware that people listening to music will often tune their actions to the rhythms of the music, which can be a problem if the rhythm played does not work well with the rhythms of the actions they are taking in the program. There will be more about this in Section III, Chapter 18.

Song differs from music in how much attention it will draw from the listener and how much meaning it will communicate.

Music affects feelings. Song as a form of poetic writing also communicates aspects of what's going on inside. These two work extremely well together.

Song is one of the most effective methods of creating memorable awareness. All cultures use song as a way to teach basic ideas and impart a sense of importance to what the songs are about.

Music and poetry blended so thoroughly and so early in humanity that many people don't see them as two distinct arts.

But they are. Composition of music and lyric writing are two distinct skills.

Singing is related to instrument playing but also involves language, and language always carries cultural attitudes with it. There is a truism that poetry is what is lost in translation, and this is doubly true for lyrics.

Music can be transported from one culture to another through acclimatization. Song has to be translated if it has to cross language barriers and cultural distinctions.

Songs are good for dense, moving communication. They have a heavy dose of information in them because of strong associations. But they can make it more difficult for users to focus on much else.

If you put a song into your program, be aware that your audience may stop what they are doing in order to listen. Songs are highly effective ways of transmitting lore, but are not as good as music if you wish your audience to be taking action at the same time.

Sculpture and 3D Printing

Most appearances of sculpture in software have been still images or videos taken of the sculptures. Multiple pictures can be presented of a sculpture from different perspectives to show various views. Video makes it possible to walk around sculptures and show the depth, complexity and change of shading. This use of videos also enables the creation of virtual forms of sculptures to put in animation.

It's pretty rare to directly commission physical sculpture to be put into software. People do, however, record sculpture, furniture, and architecture as sources for 3D models.

But that's also fading away. These days things are moving in the opposite direction, with 3D modeling producing real-world sculpture by 3D printing.

At the current time the art and software of 3D printing is only beginning to mature. There is also a great deal of experimentation in materials. At this time it would be unwise to speculate too much on where this new art will go or how it will influence the software that makes it possible.

We are in the midst of a change in sculpture itself. Working with sculptors these days is a learning process for all involved. The art and the arts that make the art possible are rapidly changing.

If you are involved in this from any side do not neglect any possibility that looks interesting to you. The space is wide and getting wider.

Animation and Video Making

Had I written this book 40 years ago, I would have talked about the emergence of CGI in special effects. Thirty years ago, I would have discussed computer animation as a new field for professionals using expensive software on highly advanced workstations and would have talked about a friend of mine studying animation in art school. I would also have mentioned the ready availability of video cameras for dedicated hobbyists. Twenty years ago, I would have discussed advances in games. Ten years ago, I would have talked about video upload sites.

Today, computer animation is a skilled art but no longer one that is prohibitively expensive to get into. Video cameras are part of nearly every phone sold. Video upload sites are easy to access. Video editing software is easily accessible and learnable. Today's computers stream, download, store and play movie-quality videos all the time.

The highest-end videomaking and animation are still the province of major movie studios and the smaller shops they commission, but the speed at which cutting edge video and animation technology becomes regularly available to anyone interested is measured in single digits of years.

Children are learning to animate and to make their own videos. That is what has happened to these formerly esoteric arts in which select studios had their own proprietary software and special effects.

It might seem that there isn't much to discuss about how to integrate animation and video into software. In a lot of cases, it's drag-and-drop.

But the same basic problems that existed 40 years ago still exist. Animation can still push the limits of processors and video cards and videos can still take user attention away from what else they are doing.

Animation in your programs can do a lot for the interface, but it can also distract users. Effects in games can look great the first few times but can wear on the user if they happen too often.

And, of course, animation and SFX have the problem that all emphasis is no emphasis. If everything is an overlay of special effects, nothing special is happening.

Use the tools for what you need them for. Don't let them take away from what you are doing.

Videomaking covers everything from snippets of a few seconds to multihour presentations. Very short clips can pop up in social media or in pop-ups anywhere, and they can annoy and distract. Extremely long videos can pull people in for long commitments of time without necessarily giving much interesting.

This rapidly burgeoning field is growing and has become easy to use. We have reached the stage of the art where the question is no longer can we do this, but when should we do this. What range of emphasis, what stretch of time is needed for the particular effect we're after?

Programmers and animators/videomakers share a fundamental awareness. Video relies on persistence. The art itself emerged from the discovery that showing images rapidly to the eye created a sense of motion. Flipbooks gave rise to nickelodeons which gave rise to movies and so on.

We can make tools for them and incorporate their work in ours. Our manipulation of persistence and their creation of motion from persistence makes our arts kindred. And like all kindred we can simultaneously get along well and fight pointlessly. Working together takes a special level of care, because of how deeply we share.

Voice Acting and Audio Recording

Voice acting evolved in the golden age of radio, the first era of non-visual telepresent art. This was also the era in which Foley evolved from its origins in theater.

Voice acting is the skill of using people talking in order to give an awareness of space and character.

Voice acting can be used to present character or narrative voices to listeners. Characters can either be speaking out loud and conversing with other characters or declaiming their internal monologue. Narrative voices can describe environment or what is happening to characters.

Voice acting is a different skill from acting (which we'll talk about later). When properly done it's extremely effective at creating materials for listeners to create their own imaginary sense of what is happening.

Between the end of the heyday of radio and the emergence of podcasting, pure voice acting was not much regarded. Its secondary use, providing voices for animated characters, became its major use. Skilled voice actors gave voice to generations of animated characters in cartoons, full-length movies, TV shows, and advertising.

Currently, voice actors provide all of these things and more for software. They give voices for animation and provide voiceovers and narration. They provide snippets of speech to make it possible for icons and avatars to speak what they are doing and ask questions of users.

Making this work well is difficult.

People rarely notice how everyday speech is not at all like dialogue. People conversing have a lot more hesitations, broken pacing, and change of affect than they might think they do. As speakers we are used to talking this way. As listeners we are used to transforming this kind of speech into edited memory.

The skill of speaking fluidly with nuance and personality is difficult to master. Working with one or more voice actors to present scripts in your software, whether full-on narrations, voicing animations, or short bits of help or suggestions, can give life to what you are presenting.

If your software requires both animation and voice acting, the voice acting should come first. Animators put in facial motions to match what is spoken. One of the reasons dubbing from one language to another is so difficult is that it's hard to write a dub that would move a face and lips in a way like the motions in the original language.

Audio recording has returned to popularity through the twin arts of podcasting and audio book recording. Both of these are related to voice acting but diverge from it, each into their own way.

Podcasting is usually monologue or small group dialogue about various topics.

Podcasts may be fully scripted, fully improvised, or mixed (improvisation with preselected subjects and talking points). Some podcasts are stories or essays or debates. Some are like old-fashioned radio dramas or soap operas, with long, evolving tales where strange or amusing things happen.

Audio books are somebody reading a book out loud, preferably a good voice actor who can do various voices for characters and has an interesting and involving delivery. A normal-length novel will take about ten hours of reading on average.

Even if you do not put podcasts or audiobooks into your software, you should be aware that many people like to have one of these going while they are doing other things. If your software requires that your users be listening to its voices, music and sound effects, then they will not be able to also listen to podcasts, audiobooks or their own music. Many people find that what they do is easier if they have control over their audio experiences. There will be more about this in Section III, Chapter 18.

If possible, it's a good idea to make it possible to use your software without audio tracks playing. Many of your users may have other things they wish to devote their hearing to.

Sometimes the way to involve an art is to not involve it.

Acting, Dance, and Puppetry

Acting presents characters using a wide variety of ancillary activities. Actors' bodies are in motion while talking, wearing costumes, using props, stage fighting, dancing, singing, embracing, hand-holding, having expressions of all kinds, looking toward, looking away, looking askance, and so on.

There's a lot going on, and a lot of other artists that can help actors do their work: Scriptwriters, prop makers, costumers, directors, choreographers, fight choreographers, set designers, set makers, etc.

Filmed acting involves cinematography, editing, and post-production work. There's a reason movie and TV credits lists are so long.

All of these arts can be used to create live-action videos which can be employed directly in one's software.

There are also ways to derive data from the actions of actors.

Motion capture is at the moment the standard method.

Motion capture can gather in body and facial motion data and use it to move and shape models. It lets the art of acting pass through into animated bodies and worlds. It makes the animated body into a costume for the human body and then removes the human body entirely.

Motion capture is related to both dancing that shows bodies in motion, and puppetry that transforms hidden human motion into the visible motion of something else.

Motion capture moves real-world motion into virtual spaces by gathering data and interpolating movements from it. The process relies on the complex information of a human in motion who is acting and dancing.

Data are abstracted and used to create a smoothed-out, somewhat stylized version of how humans move.

The stylization is visible when motion-capture-controlled figures are put into the same scenes as actual real-world bodies in motion.

Data next to information creates a contrast.

There is a tendency to treat this contrast as a problem and try to make animation hyper-realistic.

But that's only one art style. The placement of unrealistic figures in artistic motion in the same spaces as actual humans can produce wonderful art, as seen in nearly all Muppet movies and TV shows.

Making the contrast of life and mocap part of the art is still being explored as part of the rapid evolution of animation.

Hardware and Robotics

The last art we're going to look at is arguably the first art coding must deal with, the art of the hardware the code runs on.

We haven't looked too hard at hardware because it evolves and spreads faster than software.

At one point in the early years of the 21st century I asked a class of middle-schoolers how many computers they had in their houses. This was to make them aware of how many household devices that had CPUs and memory in them and were controlled by code. That number averaged around ten.

In the two decades since I gave that talk, that number has increased dramatically. So many things are now controlled by computer and interact wirelessly with other similar devices that the phrase "internet of things" has come into vogue, if not everyday usage.

Each of these devices is standing at the pivot between the information environment and the data coding. Each needs to be able to go back and forth between them to do what it is made for.

Most of these devices are software-enabled versions of tools that preexisted, like app-controllable thermostats and programmable microwave ovens. Some are far more complex, such as security cameras with facial-recognition software. Then there are robots which have to move and work in the information-dense world.

Programming for these devices has revealed layers of unexpected difficulty that in hindsight are obvious (Missing the retroactively obvious, by the way, is a commonality between science fiction and technology.)

For example, a number of facial-recognition programs carry the biases of their programmers. The people who wrote the software and trained the systems put their ways of looking at people into the data. They didn't realize that because they look more closely and carefully at one group of people than at others, they would create disparate levels of accuracy between those people they stereotyped and those they didn't.

We are dealing here with the most output-centered parts of programming, where what is brought into reality must interact with the information-dense world and become part of the entanglement. Unforeseen conflicts are foreseeable.

How much these problems can be corrected in operation depends on the level of autonomy of the devices.

Robots have been programmed with varying levels of autonomy because of how distant they are from operators and how rapidly they need to react to situations.

Robot probes far out in space need to be able to receive and process instructions for what to focus on. Most of their actions had to be created and tested in artificial environments on Earth. They needed to be programmed and debugged to do a specific set of tasks in an information-dense environment. Their overall success rate is good because of how carefully this is done.

Drones are mostly teleguided rather than being fully autonomous. The task of guiding them through the information-heavy world is given over mostly to humans because the actions they need to take are not easily reduceable to simple anticipatable tasks.

Robot makers are currently attempting to create robots that can navigate through possible actions that have a risk of harming people in a way that is safe for the people.

It is an interesting and concerning question as to whether this problem is actually solvable. A large number of solutions are currently being tested. None with a great deal of success.

It is worth noting that humans in such situations still often react with fight, flight, freeze or fawn. Humans can learn to ignore a large number of factors in such a situation and carry out a single practiced response. So we either do one of the four fs or we do what a simple robot would do.

It is possible that the information complexity and speed of reaction needed to solve some of the real-world danger problems we encounter cannot be solved with data. Our list of possibilities is not better than that of a non-sapient creature or pre-programmed machine. It may be that sapience does not actually help with danger situations. Sapience may shine in avoidance of danger. Sapient mind may need time beforehand to sculpt the situation to advantage.

I could well be wrong. A human solution or a computer solution may be found. But we also need to consider the possibility that we as information-processing beings with data-processing assistants will always eventually run into problems with too much information where our best bets in this information-dense, probabilistic universe is to try something that might work, knowing that harmful outcomes cannot be completely avoided.

And then we need to face the question of whether we should create robots that will certainly carry our personal biases as to how to react with them.

Arts Integral, Unknown, and Unregarded

This chapter is a snapshot of known and regarded arts that at this time have a fair amount of direct interaction with the art of programming.

It does not cover those arts that are so firmly entangled with aspects of programming that they cannot be separated.

Game design is so much a part of game programming that the designers and the coders must work together and in a real sense are sharing the res of the program. Deeply integrated arts like this need books that dig deeper into that integration than we can cover here.

Everything humans do is art. Everything humans do recurrently can be made into an artform. As we evolve our tools and our ways, we turn more things into materials and more actions into techniques. In so doing we expand the space of arts.

There are unknown arts as yet unmade and there are arts practiced in various places that have not been elsewhere perceived as arts.

As time goes on and the unknown evolves into the known and the unregarded arts are noticed, different arts will enter the picture, creating new kinds of art-to-art interactions.

The arts that have already been laid out will change how they work with coding. And coding itself will evolve, changing how it is of use to and how it makes use of other arts.

In all these cases, the same set of concerns will recur.

How do we use this art in our software?
How can software help artists of this art?
How can a mind that practices programming and another art blend them together?
How can different artists blend their arts with programming?
How can this art be part of these people's lives?

Personal Exercises

1. Take any non-programming art you have some skill in. Consider how you might incorporate some of that art in your code. Look at the idea from the programming perspective and from the perspective of the other art.

Testable Exercises

1. Take any art you do not practice and consider something makeable with that art that you would like to incorporate in your programming. Write down a description of what you would like. Make a guess about how long it would take an artist to create what you've estimated and what price they would charge. Look online for artists who do that kind of work and see how long they take to do such things.

   Do not contact the artists! Look up what they have posted about their work. Compare your time and cost estimates with what the artists have said.

2. Find a transcript of people talking and mark it up to remove all the breaks, hesitations, and non-contributory words to see what it would sound like as dialogue.

3. Find three video clips of CGI one contemporary, one from 10 years ago, and one from 20 years ago. Write out differences in capability that you notice.

4. Use a program that shows video or plays music while you are using it. Detail how well these work with your user experience and how they clash with it.

Section III

Foreground
Artist's Life

16

What Is Needed?

Looking at Humanity through Binocular Vision of Art and STEM

The purpose of art is to make what people need. This begs the question of how to determine what is needed that we can make.

There is, of course, no one practice to determine this. So we're going to look at metapractices that are likely to be adaptable to the purpose.

We'll start with considering where we are. This is one of those vital metapractices that people often neglect. It's easy to get lost and distracted, or to assume that all places are somehow alike, or to assume that each place is so different from all others that there's no comparison.

Where are we now? We're two thirds of the way through a book on programming as art. Looking around, we can see that we've elaborated aspects of both STEM and art and how they are one entangled awareness. So where we are is a place that makes and combines tools of perspective.

We're going to push this a bit in a metaphor, wherein we have two distinct lenses for our eyes that combine to produce binocular vision.

Each of STEM and art perceive different, mutually entangled aspects of the world and humanity.

If we bring together those visions, we will bring together the aspects of the world and human need they each illuminate into a view with greater depth perception.

STEM presents the currently measurable universe with the current evolving theories by which it is understood and the practical applications for capability that can be made using this knowledge and understanding. Together these give us a sense of what we can affect with our current evolving technology and what it takes to affect those parts of it.

STEM tends toward high specialization so that what a person doing a STEM job knows and knows how to affect might cover only a narrow space of reality, human need, and experience.

This leads to two common STEM errors, the error of seeing everything in terms of one's own field, and the error of assuming that everyone else's fields are either incomprehensible or trivial.

Both of these are solvable by understanding enough about other fields to be able to converse with people in those fields and to be able to quickly search out what is currently known, theorized and under study. Doing this properly leads to the ability to respect others and see that their fields are as deep and interesting as one's own.

It takes some work and time to learn more than one's specialty, but it has become relatively easy to get enough comprehension to be able to listen. There are reliable online sources that cover many aspects of fields. And it is possible to formulate questions to help find out what is being worked on and what isn't.

The main thing is that one does not need to be able to do the work of a field to gain some understanding of what the people in the field are doing.

It's also important to be aware that whatever one may have learned the last time one looked into a field (which might be as long ago as grade school) is likely to be well behind the curve.

The tendency for STEM people to overfocus on their own fields and not see that other fields are equally complex and deep is not automatically dealt with by getting a sense of other fields. The "Now I've got it" eureka sense needs to be tamped down. Seeing a part is not seeing the whole, and benefitting from a bit of understanding is not the same thing as being an expert.

Being aware that one's knowledge and understanding will always need updating, and having the competence to do that updating is vital to keeping one's STEM eye open.

The art eye can see what people are focusing on, what they are doing, what they are enjoying, and what they aren't. It can perceive what art spaces are occupied and what materials, techniques, and tools are being used in those spaces. It is also good at awareness of social action, interaction, and pressure.

To turn the art eye toward something, pay attention to some aspect of the clothes, music, food, dances, videos, pictures, games, and so on that are popular right now, and which are niche interests. Also, take a look at what older art styles are making a retro comeback.

Take one step forward from both the STEM eye and the art eye and put the visions together, using such understanding as you've developed from your own awareness and from the earlier parts of this book.

Together STEM and art can perceive what is currently perceived and used, how it is perceived and used, who is perceiving and using it, what its past looked like, and what its current visions of future possibilities are focused on.

This kind of awareness can help one survey humanity in whole or in part and derive an overview that can help the STEM/artist's mind perceive what problems exist and what art might be made to help with them.

Seeing the Problems That Are Most Visible to You

The problems that directly impact our lives are not always the ones that we can most easily see.

The human ability to acclimate and normalize causes us to miss a great many things that make our lives harder. Deacclimatizing and denormalization takes a great deal of res work and often a very long time to process. Outside perspectives can help, assuming the people with those perspectives are listening and paying attention to others rather than prescribing cure-alls.

The problems that we are most likely to notice are those that recur often enough that we recognize the recurrence but not so often that we treat them as natural parts of the environment in which we live.

The problems we are second most likely to notice are the problems we see that trip other people up. This can also take some denormalization. There are a number of social harms that dismiss or mock the difficulties people have. It's crucial to remove those attitudes in order to see real problems that one might take part in solving.

It may be that something you have been acclimated to becomes suddenly visible as a problem that needs solving.

Often this means that the back of your mind has been working on a problem well before the front of mind is even aware that it exists.

Back-of-mind awareness often causes front-of-mind annoyance and possibly hypersensitivity. If something has been bugging you, take a closer look and you may see the problem within it.

Listening for Problems That Others Are Having

The space of available problems expands widely if one starts listening to other people talk about their lives, work, hobbies, arts, and so on.

Some of these will show up in their complaints, but not as many as you might think. You're likely to find many problems in actions that harm them but that they've normalized. Finding out that someone spends several hours a day on tasks that could be automated or sped up or eased with an appropriate tool is as fertile a ground for problem-solving as are the places they audibly wish were different.

Difficulties can be mined out of complaining, although it's complicated.

If you know someone well enough to know their complaining patterns, you might be able to ask questions that lead to what they are actually having trouble with.

But not everyone appreciates this, and some people don't wish to have their sources of complaints removed. You still might work to remove those causes, but they may not be happy with you.

It is possible to expand the circle of listening beyond the space of people you know into the wider world, to read stories and reports, take part in online conversation and so on.

This takes care and skill. At all times there are large-scale shared complaints that are inaccurate and smaller-scale needs that are real. Finding the small-scale needs involves delving into various groups that one is usually not a member of. Understanding other people's perspectives on their problems involves trying to understand their view of things. This takes a lot of listening before even asking an initial question.

Developing the skills to listen to other people talk about lives unlike yours and problems that may or may not resemble ones you know is worth the time and care, but it will not come easily.

The first and most basic of these skills is understanding whether or not one's presence is welcome in someone else's space.

As mentioned before, STEM and art both, unfortunately, teach the idea that the world and humanity are ours to observe, analyze, and fix as we think we should. Both of the lenses we are using tend to erase the boundaries people draw around themselves and their cultures.

Listening as one human respecting the lives and ways of other humans is the first step and the last step to not falling into this error.

Are They Really Problems?

Having found what seem to be problems, the next question is, is it really a problem?

A person may be frustrated that a thing cannot be done simply, that there is no button they can push to get what you need. They may be annoyed that learning is necessary in order to do something.

This might conceal a real problem, in that no currently available learning method works for this person. It might also be true that a problem could have a pushbutton solution if it required no art. But many times this is not a real problem. Most human activities really do take work to learn.

There is also a kind of annoyance that is not actually a problem. If a kind of art you like has fallen out of fashion or is not done as you like it where you now are, you may be frustrated at not being able to get it. People also get annoyed that social customs change, that what is acceptable for people to do changes as societies evolve.

These changes may or may not be real problems depending on whether and how they increase or decrease fulfillment of human need.

Here we are steering between mountains of discourse (a great deal of it violent). There are people who object to new ways arising and old ways passing away. There are people who object when old ways continue to evolve.

There is also the recurrent error wherein people think that the world would be better if everyone thought and did the same things.

I tend to take an evolutionary (but not evolutionary psychology) perspective that monoculture is unwise. My view is that humanity thrives on individual and group diversity because that optimizes the possibilities that people have access to and because it derives from the fact of human uniqueness within the framework of human commonality.

So I, personally, do not trust any assertion that monoculture is a good idea. Nor do I think that people having and making art they enjoy whether it accords with a particular culture's attitudes is an actual problem.

Finding the Problem within the Problem

Assuming you have identified an actual honest problem, you have to determine if you can find a solvable problem within the identified problem before you can deal with how to solve it.

Apparent problems are surface expressions, structural difficulties arising usually from infrastructural considerations. Real problems are mostly infrastructural. But what infrastructure you'll be working on will depend on the art you will use to solve it.

Assuming you are going to be creating your solution through programming, then the interior problem is the core one that you will solve by coding.

If you are going to solve it by other means you might need to seek for a different interior problem than the one you would look for to solve by coding.

If you think a multiple-artist solution is needed, you may need to discuss with others what the interior problem or problems are and how to coordinate the solutions.

The important thing to realize here is that finding the interior problem is relative to how you see problems and solutions. A fundamental aspect of art as a source of solutions is using the artist's sense of that art to find the problems that speak to you as solvable and hint at how they can be solved.

You will likely need to do research and other res work to be able to find the solvable interior problem. The result of that work is often the first form of the res that will guide you through the artmaking.

One thing to be aware of is that sometimes a standard method will present itself as a way to solve a surface problem.

Be cautious when this happens. Standard surface methods don't work well for finding interior problems. It's not uncommon to spend a lot of time and effort trying to impose methods that are standardized but do not fit because they presume an infrastructure different from the one you are actually dealing with.

Examine the information infrastructure to see what is actually happening. See if it really corresponds to the assumptions made by a standard solution. If not, then you may need to discard the standard solution.

Once you have the inner problem, try to solve it as you would solve it using the programming/art methods that work for you. If that doesn't work, go back to consider the outer problem again and reexamine it to see if you can find a different interior problem that might be solvable as you solve things.

Who Does Your Solution Solve the Problem For

Before spending the hours/days/weeks/months/years it takes to embody your solution in freshly debugged operational code (or another art form), consider who your solution will work for.

What does it require of the users? What mental and physical tasks will they need to be able do in order to use it? How will it fit into and interact with any other relevant tools or toys in their life, work and play?

It's a good idea to examine your solutions from this perspective and see how and if the space of people it can work for can be expanded.

A tool of very simple use, such as a screwdriver, can be widened in applicability by giving it a few options and perhaps a versatile handle.

A complex tool, such as a word processor, can be broadened in use by adding generally useful features and options for specialized uses.

Be careful how you add widened possibilities. It's not good to drown out the initial problem you were solving by making the program too noisily feature-filled.

All solutions need to be human solutions. The solutions may be deeply infrastructural, so that few humans are aware that they exist or that hidden problems have been solved on a level they know little or nothing about. But nevertheless, the entanglement of information and the complexities of reality and life can make even something as far away from human thinking as, say, semiconductors create revolutions in how humans live, work, play, and store.

Your most hidden work may stretch out to your widest audience, even if they never know what you've done or whose lives you have changed.

Finding More Problems as You Go Along

As you devote the hours/days/weeks/months/years it takes to embody your solution, you are likely to discover other problems related to your initial motivation.

This is a problem with materials. Perhaps the language is not well suited for particular implementation methods.

This is a problem with techniques. You find your recursive invocation loop is a memory hog.

This problem is user-based. Your creation may require someone to live-monitor a process for multiple days without sleep or a forest will catch fire. See the history of charcoal burning for this real-life problem.

You discover something fundamental to the universe. The 3 degrees Kelvin microwave background radiation left over from the big bang was discovered because the users of radio telescopes couldn't clear away an annoying disruption to their signals.

This problem interests you. The problem fits with your awareness of problems and you can make use of solving it in more ways than one. You may solve this instance of the problem and extract this problem for deeper exploration later. Or you might spend the time and effort to solve the deeper problem and use that to solve this single instance once you've taken care of the broader problem.

This problem interests someone else. This might not be your thing, but it might be of particular interest to someone you know, in which case you might hand the problem over to them and see what they think.

This problem is an unending fractal of complexity but it's not fundamental or interesting. Back up and find another path.

Helping Your Works Be Useful beyond Your Vision

There is a persistent myth that the first person to invent or discover something is the person who knows it best.

Often the opposite is true. The original insight into something finds the beginning, the way into unexplored territory, but does not map the whole space out.

Newton, Leibniz, Reimann and others invented calculus. None of them were anywhere near as good at knowing what can be done with it or solving problems with it as someone these days who passes an introductory calculus class.

Each person's vision and awareness of their work extends only so far. Multiple people working from that work extend the space of use of the work, creating a field that can be learned and expanded further.

The problems you solve and the art/tools you make can give ideas to other people and grow their uses beyond your personal vision.

Making it easier for others to expand upon your work can be done with documentation, with input/output elements that reveal the interior processes, with online discussions and bug reporting facilities and so on.

It can also, oddly enough, be done by patenting, which we'll talk about in Section III, Chapter 23.

And, of course, it can be done by someone looking at the way you've solved things and finding their own problems therein.

Personal Exercises

1. Consider what current art trends you are interested in and which you don't care for. Try to put yourself into the place of someone with the opposite view of your interests.

2. Look at any STEM field you have a pretty good knowledge of. What aspects of it do people outside the field generally know about? What areas do you wish they understood and which frustrate you because they don't?

3. After doing personal practice 2, consider a STEM area that you know only somewhat and see if you can find out what parts of that field the people in it wished that people outside knew.

4. Talk to people you know about any difficulties they may be having and that they are willing to talk about. Make it clear that this is an exercise, but that you are honestly interested. Don't push. Listen and make notes. Ask questions. Do not make any suggestions. Just listen. Thank them.

Testable Exercises

1. Take a field of art that is currently undergoing a retro revival. Compare the original works to the retro. Describe the differences.

2. After doing personal practice 3, research the aspects of the other field that insiders wish outsiders understood. Write down any "Now I get it" moments you experience, and see how much more there is to it than your eureka moment.

3. Take something that you do not have difficulty with but that many other people do. Document social attitudes toward this difficulty. See what solutions have been offered, including dismissal and disregard of the people who don't get it.

17

What Do You Enjoy Making?

Playing with Ideas

The perspective generated by solving problems based on need does not on its own answer the question of what problems are a good idea for you personally to be solving.

Yes, work based on need is real and necessary, but it's better if the work you do is also enjoyable for you. Art that wearies the mind is rarely as expansive, nuanced, complex and inspiring as art that the artist enjoys doing.

We glossed over this in the previous chapter, but now let's look directly at the question of what happens when a mind plays with the ideas that arise from and go into the problems and solutions we are working to solve.

Enjoyable ideas tend to have a mobility and a flexibility to them.

How those qualities will manifest varies from mind to mind.

They may be light and easy to move. They may cause other thoughts to organize around them. They may flash and shine. They may bounce around like cartoon characters. They may simply make other ideas make sense, or be easy to connect other ideas to.

Regardless of internal manifestation, the act of playing with those kinds of ideas will get the mind flowing through wider spaces of possibilities.

These possibilities might manifest as running through lists of ideas, or coming to sudden clear conclusions, or exploring territories, imagining events or conversing with characters.

Whatever the form, the playing feels good and the good feelings widen the region of consideration.

Some ideas may feel heavy, inert and unchangeable until seen from some particular point of view or within some convivial context, at which point they become mobile and flexible.

Some mobile ideas don't generate a sense of lightness. Their enjoyability moves implacably like freight trains toward where the mind needs them.

Sometimes the mobility and enjoyability of ideas makes them become large, deep and gravitic, so that they attract other ideas to them and form worlds that grow ecosystems of ideas that the mind can make use of for many projects and interests.

There is no one single experience of enjoyment uniform to humanity or even uniform to a single mind.

Enjoyment is a part of understanding, inchoate and working behind the scenes even when it creates visible manifestations in the front of mind. Back of mind has a thousand hands to hold up myriads of manifest forms to play and interact together.

DOI: 10.1201/9781003459866-20

Playing with Implementation

There is no not playing with implementation. The feedback between res and dictum is mind playing with possibility and actuality until what can be and what happens to be are so entangled that art exists and tools work.

Even someone who works using a fixed methodology trying to reach an exact formulation is playing inside their minds and playing with what they make.

The question is not whether someone is playing when they blend thought and reality into humanity. The question is whether they are playing in a way that works for them and their audiences.

There is a diversity of possible implementations for any res. Many more possibilities can be made real than have never been tried.

Some people explore this space by testing multitudes of materials, or experimenting with techniques from many schools of art. Other people explore the space by working in small regions of form or limited palettes of materials.

In cooking I work with limited palettes because my tastes are narrow, but I study techniques from cuisines all over the world. In programming I pick up languages and play with them and then seek others and make up techniques as I need them. In writing I always put some poetry in my prose and I work on the principle that writers need to know everything even though they can't actually do this.

All of these personal expressions of my enjoyment and ways of playing with ideas have shown up and will continue to show up in this book. The fact that I am writing this with a conversational tone is also my personal preference. The fact that I can write a textbook with this tone is the result of a recent cultural change in how people read and what their expectations are.

This cultural change has been caused by online social networks.

The distant abstracted author creating textbooks with a tone of authority goes at least as far back as Aristotle. Yes, he wrote in a grumpy voice, but he wasn't being conversationally grumpy. He was pushing his ideas as the only true knowledge and dismissing the ideas of all others.

The more convivial protocol of chattiness caused by social networking has changed audiencing so that I can write this with a clearer authorial voice with less authorial looming. My preferred form of writing and your expectations in reading have made this dictum possible.

The important thing is that the dictum be an implementation of the res that works from the artist and for the intended audience.

Any given res can manifest in a wide variety of forms. A person who works in a narrow space has as much room for play as someone who sees vast expanses of possibilities and seeks to know all forms possible in their art.

The important thing is to find the spaces of implementation that work for you and that you can make work. As your art evolves those spaces will change. How they change is as unpredictable as the creations of any unique mind.

Enjoying the Making

There is a distinction between enjoying the actions of making art and enjoying having made the art.

There is an unfortunate myth of the suffering artist, that the making itself cannot be enjoyed.

The Romantic era in art history glorified the concept that the artist is giving up mere human happiness for the sake of expressing some ineffable truth.

This is nonsense. Human happiness is an ineffable truth.

Seriously, no one benefits from this pile of melodrama.

It's bad for artists because they don't try to find or craft praxes that help them make art in ways they enjoy.

It's bad for the family and friends of artists because the myth puts them in the position of attendants on tragic genius instead of humans living with other humans.

And it's bad for human society because it fosters alienation between art and humanity, which is alienation between humanity and humanity.

We'll dig more into what to do about these larger-scale social problems in Section III, Chapters 18 and 21. For now, let's address the personal questions of what can be done to enjoy the making.

First, find what you like to do.

If the process itself is focused and stressful, but you like the good outcomes, change what you see as outcomes. Look at making each part of the project work as an ending that you can look back from and enjoy. That way you can enjoy each day's work.

If you enjoy the doing, but not the having done, then pay heed to the process and let others take care of seeing how well things came out.

If you enjoy the coding itself but not the debugging (a very common arrangement), find a development environment with the least painful debugging tools. Find the rhythm between coding and debugging that works best for you. If you need to debug each chunk of code after you do it, then do it that way. If you need to code large sections of a whole project before doing any debugging, then do it that way.

If debugging makes you twitchy, try to arrange time to yourself to make it work and take breaks for human interaction or quiet recovery. Don't harm yourself by forcing your way through frustration.

If debugging is a game to you and you enjoy it, then remember game-life balance. Don't lose track of all human interaction while you're finding the mistakes and eliminating them.

The main metapractice for all the parts of work is not to go so far into the doing that you enjoy that you lose track of life, and not to let the parts you don't enjoy get so far into you that you lose track of life.

There is a kind of enjoyment that comes from getting through the parts of work that you don't like. It doesn't feel like fun, but it is a gleaning of energy from making your work work for others.

Human enjoyment requires an awareness of one's humanity, which is also what is needed to make art by and for humans.

Enjoying Watching the Enjoyment of Others

Some artists enjoy hearing audience reactions to their work. Others do not.

Helpfully, there is an enjoyment that can be had whether one likes hearing praise of one's work or not.

Watching someone you like enjoying an art that works for them can be its own joy, even if that art does not appeal to you.

This echolocated enjoyment can be made greater by the awareness that the art they are enjoying is art you made, and therefore you know a bit more now about how to make art that makes other people happy.

This kind of echoed happiness can be learned relative to one art and applied in others. If one cooks for other people, it's relatively easy to enjoy their enjoyment of the meals or confections one makes. The same applies to singing, playing instruments, telling jokes, storytelling, reading aloud, and some kinds of shared gameplay. Arts with immediacy of interaction rely on an entanglement between artist awareness and audience response.

Learning to gather that awareness of other people's enjoyment and craft your own enjoyment from it can be extended to other arts where there is more of a time gap between making the art and the audience perceiving it. Drawing, painting, sculpture, writing, cartooning, and coding all fall into this category.

The crucial element here is to focus on the reaction to the art, not on the art itself. Pay attention to the actions and reactions of the audience. Seeing appreciation, illumination, thoughtfulness, contentment, and excitement can spark personal happiness in the perceiver.

This praxis is not going to work for everyone. If it does work for you, it can remove a good deal of the sense of alienation many artists feel toward their own work and towards praise of that work.

What Work Energizes You and What Work Drains You?

One aspect of enjoyment is whether what one is doing gives one energy to do more.

This is not the whole of enjoyment. There are things done that take a great deal of energy which feel good or are inspiring in the doing of them but which leave one exhausted afterwards.

Such processes may be needful for work and play, but cannot be sustained.

Praxes that give energy can be kept going for long stretches of time, during which body and mind can both be refreshed while exhaustion can dissipate as energy supplies are refilled. Identifying the parts of life that energize and sustain is vital to determining how you should commit your time.

It's also vital to recognize what praxes leave you exhausted and incapable of doing much of anything.

The three aspects of this measurement of energy change are the praxis itself, what the praxis is being applied to, and who it is being made for.

The last of these is the consideration most people ignore.

There is a difference between making an artwork for a person one knows well and making art for people one does not know.

Either of these can be exciting or draining. There are people who have great difficulty doing art for people they care most about, but can produce great work for people they perceive at a distance. This is a fact of human uniqueness.

Please note that there is a difference between doing art for people you are close to and doing art for people you know but who tend to be jerks to you about your work. There is no point in making art for people who have no interest in it or use your work to tear you down. Whether those people are your family or random strangers on the internet, they are not your audience.

With praxes that overall energize you, it can be beneficial to examine the parts of the praxis to see what aspects energize, which are kind of meh, and which are draining.

One may find that small changes in process can remove the draining aspects.

For example, one may enjoy the act of coding, but one might need glasses or contact lenses for long-term staring at a screen. Many of us who need prescription eyewear have computer glasses which are about half the strength of our regular glasses. Coding or writing without these is a strain, but with them it's just fine.

What the praxis is being applied to is the subject matter of an art being practiced. If what you are working on is unpleasant or painful to think about, then an otherwise energizing praxis can be wearying. A person can enjoy walking, but what environment they are walking through can change the experience dramatically.

The energizing and draining qualities of a praxis can also change depending on one's capabilities in the art.

Many writers get into the field because they were energized by things they read and the works of particular authors.

But learning to write professionally involves a lot of examining, critiquing and vivisecting your own and other people's texts.

For many of us, this changes the act of reading from an energizing one to a draining one. We may still enjoy much of what we're reading, but we're also seeing how the writer did what they did and considering alternatives. The energy creator has become an energy demander.

This can also happen with coding. Dealing with messed-up software is annoying and can drain as well as inspire us to make new or better software.

There's one last crucial element in this. Praxes are not independent.

What you do when and what you did just before doing it and what you are going to do afterwards all affect whether a particular praxis is energizing or draining.

To benefit from enjoyable praxis, you will need to arrange other elements of your life to fit the needs of that praxis. There will be more about this in later chapters in Section III.

What Work Creates New Ideas or Inspires Research?

Energy is not the only consideration of beneficial praxis. There's also matter – subject matter, that is.

Minds need to have ideas to play with. The work one is doing and the ways in which that work connect to the too-much-information of reality are materials from which our minds make new ideas and possibilities – but only if we find them interesting.

Everyone has their own interests and ways of looking at the world to find what interests them.

If the art one is making is not interesting, or if it produces only a litany of the same ideas over and over, the mind can find itself grinding down until eventually the artmaking stops. The back of mind ceases being involved because there's nothing for it to make res from.

Art that inspires new ideas or questions that you might wish to research is likely to be good art to be working on. Art where the flow branches and connects elsewhere gives the mind more paths to follow and gather in.

Doing this does not always require that one stop working on art that seems uninteresting. It may only be necessary to pause and consider the art from different directions to find a perspective in which it is interesting.

Questioning the whats and whys of something, looking into the underlying assumptions and seeing if there are more possibilities or directions may get your mind going.

Every art that people make, every idea that has been explored has assumptions in it. Someone not interested in the subject as it's usually worked upon can generate outside questions and examine it from outside directions.

Be aware that it's harder to find new outside perspectives than you might think. A lot of outside views rely on ignorance of what's going on inside. Don't assume that the questions stirring your mind will revolutionize a field of endeavor, although even if they don't, those questions can lead you to less-regarded aspects of things that you will find beneficial to pursue for your own work and interests.

Whether you can or can't find byways of illumination in what you are working on, you will need to primarily do art on subjects that do inspire you.

Art that has no interest for you is art you will not do well.

Art that stirs your mind is art that you can stir the minds of others with.

Rare and Unheard Perspectives: Opening Spaces

Normalization can be a repressor of enjoyment and inspiration.

There are a variety of subtle and unsubtle cultural pressures to make one's work fit fashions and presumptions. These pressures often make artists hide the aspects of their minds, their experiences, and their backgrounds that are not centered in their societies.

Cultures provide ways for ideas to manifest. In effect, they clothe our back-of-mind concepts in front-of-mind forms.

The problem is that cultures vary in how much tailoring and customization they allow for various ideas. You may find that the off-the-rack forms don't work for your personal needs and tastes.

This cultural pressure combines flattening, erasure, and substitution, and it is actively harmful to both art and artist.

You may need to draw upon the less well-known and less understood aspects of your thinking, turn your awareness of your life experience, culture, upbringing and social groups in new directions and look at them in different perspectives.

This can be vital in making your art work for yourself as artist and for your audiences.

Those who share the same ways you think are likely to have an easier time interacting with your art and are more likely to get energy and ideas from it. Those who do not will be presented with new perspectives and a wider world of possibilities that they may not have known about.

We are navigating here through the largest scales of knowledge within humanity, where even our methods of navigation come from these factors and the difference between the water we flow on and the land to the sides is not easy to discern.

Humanity is firmly entangled with human uniqueness as well the cultures and histories humans have made through interaction between unique humans and evolving reality.

Attempts to flatten, erase, and substitute for this human information uniformly suppress all. The cultures that demand that everyone conform to them do so by flattening themselves. They exclude not just those outside them, but most people within them. They cannot avoid spiraling inward to an idea of humanity so small that it can be modeled by data.

No matter how hard people try to conform to normalization, they will not succeed. All they will do is create frustration, weariness and rage quitting.

Art by its nature emerges from the unique minds of artists gathering in need from the spaces they happen to exist in and the cultures they grew up in.

It is not possible to create a normalization that works for everyone. All normalizations are made from current fashions in ideas. Many of these look absurd and often cruel in retrospect. It is better to let your own art out and other people's art in than try to hold to a narrow norm.

People have always made art from their backgrounds, their own ways of thinking and their experiences, even the experiences that cultures have pretended did not exist.

Humans are uniquely composed from these. Their personal and cultural growth have always evolved from these.

Human experience is made of information. Information is deeply entangled and has traces everywhere in humanity. It cannot simply be erased or overwritten like data.

Humans and the art they make will always affect each other so long as humans exist. It is best to do so with care, awareness, attention and enjoyment, working through your unique and shared perspectives and considerations.

Personal Exercises

1. This is one of those chapters where every section is personal exercises.

   After reading the chapter, go back through each section and consider how you personally do what the section talks about.

Testable Exercises

1. Discuss any region of ideas that you used to dislike and now enjoy. Try to compare how you used to perceive/think about those ideas and how you now think about them.

2. Discuss any work process you dislike and how you deal with this. Look into or invent three alternative ways and try them. Document the results of trying them.

3. Document a practice you enjoy, making note of which parts work well for you, which you just make your way through, and which are actual drawbacks to the doing of it.

18

Working Alone

The Mind Is Bad at Caring for the Body

Working alone as an artist means being the single person using a set of materials, tools, and techniques to create individual artworks for audiences to make use of.

It's important to be aware that an artist working alone is part of the flows of need and fulfillment and therefore is always interconnected with other humans.

Artmaking connects to other people, the audience who will see it in the future, the artists who developed the techniques in the past, the providers of the materials in the present, and all other artists doing their own work at all times.

This is a useful image to prevent too much sense of isolation. But it's only an image.

Working alone is its own form of separation within the composition of humanity. Being the only mind involved in shaping the res and dictum of the work and determining what practices to do when is qualitatively different from being part of a group of people interacting to create art together.

This chapter is about what one needs to do in order to work alone.

We've covered the artmaking itself. This is all the other stuff a human mind needs to deal with while making art.

The first fact that needs to be faced is that the human mind, including your mind, is a phenomenon of human life, including your life, which involves the human body, specifically your body.

People have been complaining about bodies for a very long time. They've also been not doing what's needed to care for their bodies for that very long time.

Human self-awareness does not do a good job of bug reporting. One particular alert, such as a headache or nausea, can have hundreds of different causes. Any debugging environment that gave such a generic response (and there are such environments) would cause programmers to rage quit.

A lot of things we need to do to care for ourselves are counterintuitive (such as doing some exercise when tired from sitting too long) or require shifting of awareness from what we're attending to.

It's not uncommon to discover that one is very hungry because one has been paying attention to one's coding for seven hours straight.

Neither the back of mind nor the front of mind does a good job caring for the body.

The back of mind will tend to run roughshod over one's own needs in order to get a res made into a dictum. It will interrupt just about anything, including food, sleep, or recovery from illness to cause ideas to be given form.

DOI: 10.1201/9781003459866-21

The front of mind is susceptible to fashion and to self-image issues. It can form attachments to ideas of what would be right for one to do without regard for actual need. It can also get into a have-one-more-bite-you-liked-the-last-one-you'll-like-the-next error cycle and make it so that one does not stop doing something one has had enough of.

Fortunately (in a very odd sense of fortunately), back of mind and front of mind don't inherently work well together.

The front of mind can remember that the body needs to eat when the back of mind is trying to push for more coding or drawing or writing to be done.

The back of mind can be aware that the front of mind is just following habit or fashion and push forward an awareness that flows away from the habit.

Each can act as a check on the other.

Both sides of the mind benefit from a support network of outside observers who care for the whole person, not just their art. There will be more about this in Section III, Chapter 21.

The Mind Is Bad at Caring for the Mind

Consider the versatility of human minds:

The fact that memory is also imagination. The ability to craft and learn new praxes. The ability to make data from information and information from data. The capability of acclimation. The fact that everything humans do is art and that we live in an augmented reality. The ability of humans to learn from narratives of each other's experiences and mistakes. The ability to share and shape understanding through teaching and learning. The forming of societies and the ability to teach young humans how to live in their societies.

All of these features of human minds are also massive bug factories.

Human minds evolved. Evolution operates to create individual capability which is then tested on a large statistical scale.

Evolution does not set a good standard for self-care.

The needs of the body are not simple, but they recur and can be categorized.

The needs of the mind are far more diverse because each mind is unique and it grows in its own ways through its own experiences which it learns from uniquely.

What an individual mind needs is its own personal multidimensional space to create and explore and fill with its art in the making and the completed art of other people.

Fortunately (really this time), human minds evolved to do just that.

A mind creating and exploring a wide enough space of possibilities can, with help, eventually find or craft the praxes that can help it live the life it needs to.

But it needs an environment that is conducive to such development.

Human minds are quite capable of absorbing or developing self-destructive and self-sabotaging habits that they give priority over developing art and enjoyment.

Many minds grow in environments that produce just such anti-praxes.

Normal Is a List of Statistics, Not a Way of Life

Many cultures and subcultures create images of what people in the culture are supposed to behave like. They also create images of how people are not supposed to act.

These images are usually derived from a combination of generally recognized ways of behavior and cultural ideals contrasted with bad examples.

These images are often treated as though they were eternal concepts, with no variation in the ways people have acted and should act.

Historical studies show that these ideas of eternal behavior are never factual. Social forms and images evolve only slightly slower than the speed of fashion. Some can hold on for millennia and then disappear in a generation. Some are created in a given time and projected backwards, depicting people in the past as following the fashions of the moment.

The evolution of social praxes and the diversification of them as societies grow and spread show that there are benefits to societies with non-oppressive social expectations and customs.

The fundamental problem is that societies have an easier time creating norms than they do creating distributions, and humanity is distributed, not normalized.

Norms are individual artworks.

Distributions are a higher-scale awareness of how art can grow and change.

The more aware a society is of this, the easier time it has dealing with the inevitable entangled evolutions of how people live and the awareness of how people live.

Forced adherence to norms is harmful, as is forced ignoring of human need. Humans trying to force themselves into socially prescribed norms of body and mind do a great deal of damage to themselves and others they interact with.

Bodily Health

The body needs a lot of things. There are four that artmaking in general and programming in particular commonly interfere with: Food, sleep, medicine, and physical activity.

Programmers have a long tradition of eating really badly. Living on snacks and soda while coding is a point of pride among many programmers.

Stop this right now. Seriously. Cut it out. This is actively dangerous. And no, protein shakes and other nutrient supplements are not a substitute for eating food that you can focus on and enjoy, either alone or with other people.

Programmers are developing diabetes and high blood pressure and a bunch of other conditions that arise from treating food as fuel rather than as a complex human need.

Our bodies are not simple. Our bodies are made of cells that are not simple.

Eating, as we discussed before, is the front end of an entangled skein of complex processes that sustain the human body.

Make space and time for eating food you enjoy and that is good for your needs.

And, if you enjoy it, don't neglect the benefits of dining with other people. Humanity evolved conversing while eating. It's socially beneficial – even if it is biologically weird since we can't talk and eat simultaneously. (On the other hand, eating is one way to listen instead of talking, so maybe it does make sense.)

Not everyone benefits from sharing meals and, of course, it matters who you're sharing them with. Being alone is better than harmful company. Even so, the combination of eating and conversing that comprises dining can make life more fun and interesting for lots of humans.

Conversing while eating is a good opportunity to talk about what you are doing with your work and what you might need to be doing soon and to listen to others talk about their art and their needs.

There is also the matter of cooking. The hardest form of cooking to make interesting is to cook just for yourself. People tend to oversimplify the food they make alone and that can lead back to a diet of snacks and sugary drinks. Cooking for yourself and others gives you another art to work upon and more of an appreciation for the art you are making.

Cooking is one of the best arts to practice in order to develop an awareness of the diversity of human need and tastes. After all, the idea that taste as a concept about which there is no dispute comes from eating.

If you don't do your own cooking or you share cooking times with someone else, you can expand your audience awareness through seeing how cooking is done relative to different people's needs and enjoyments.

This is not a plea for a snack- and sweets-free diet. Far from it. However, most snacks are made of so-so ingredients and are far less satisfying than they could be. Snacks can be enjoyable. Baking and confectionery are fun arts all on their own. Good-quality chocolate, for example, is in general much less sugary and much more satisfying than most candies. (By the way, chocolate truffles are really easy to make and very satisfying if you like chocolates.)

There is a Zen saying, "When hungry, eat. When tired, sleep."

The second is much harder than it sounds. Human bodies need sleep, but it's not simple. People have to find their own sleep cycles and they may need help getting to sleep. As an insomniac, I can say that my sleep patterns were severely messed up until I was given medicines that helped me adjust over time.

Sleep needs are more individual than most people think.

Regrettably, many modern societies have imposed sleep patterns that arose from farming, factory shift work, and school schedules.

This imposed social norm has not done anyone any favors. The worst example of this may be the sleep patterns that medical students and beginning doctors work under. Sleep deprivation is endemic among them. And that's absurd for people entrusted with the care of other people.

Programmers and other tech industry workers have been given other dangerous patterns. The crunch time ethos that has programmers working multiday stretches is justified by the claim that its shows commitment and toughness. This bad praxis has infected the entire art and gotten individual programmers to treat their own sleep needs as unnecessary.

This is physically and psychologically harmful. You need to find your own working sleep schedules and try to shape your work time around them.

But beware, back-of-mind inspiration can yank people up in the middle of the night with the need to make art. If this happens to you, try not to dive directly into starting a whole work day hours early. It's better to find what the back of mind actually needs to be done and do just that.

You may only need to make a note on an idea that might otherwise disappear and then get back to sleep, or you might actually need to code until dawn, have breakfast, and then get back to sleep. Backs of minds are really bad at time and timing.

Medicine is a very complicated subject and not just because of how long it takes to learn it. For far too long medicine ran on a single model of health and tried to get everyone to fit that model. The monocultural concept of what every human life should look like started to diversify in the late 20th century and has evolved further in the early twenty-first century. But the idea of a unified healthiness and what is needful to achieve it is still prevalent.

We'll steer around the mountains of books and studies on diversity of medical needs and go straight to the relevant fact.

The secret of medicine is that while it derives from biological science, it's really an art.

Nothing matters more in medicine than the interaction and shared awareness between doctor or nurse and patient. Medical treatments need to be given with attention to the patient's actual needs and reported symptoms. Results of tests and treatments need to be listened to and discussed honestly. Feedback needs to flow in all directions.

Current medical infrastructure is trying to undermine this need in just about all directions, and it isn't always possible for doctors, nurses, and patients to get around it.

It takes art to combine the results of tests, what the patient is saying about themselves, and current medical knowledge to figure out what the patients might need. It takes care to monitor the results of implementing those needs and see if they are working. And it takes a deep sense of what human lives can be like to see that two different people might have radically different needs even if they seem to have the same symptoms.

It's hard to be a patient in these circumstances. Getting the care you need requires that doctors listen and consider rather than dismiss what you are saying about yourself.

There is the further problem of gathering useful medical information. Being wildly misinformed about health and sickness is one of humanity's oldest problems. And it has never improved. Medical myths spread through fear, desire, and resentment.

The anti-vaxx movement began when vaccines were first created and has had varying degrees of popularity. Its danger to human life is blatant and well documented, but it has never gone away.

Similarly, the relationship between fad diets and health myths mutates in form, but is always present.

The underlying biology of this art is a science, and like all sciences it evolves and discovers where it was in error. Because of this evolution, what medicine can do and what it recommends to do will evolve as new realizations, methods, tools, and materials are developed.

I'm alive four times over because of new medical awareness and treatment. I've had four very good doctors to see me through over the last 20 years. My children had a great pediatrician who helped them emerge healthy from childhood. We are lucky in those regards.

There is no simple awareness for medicine. No "When hungry, eat." But there is the wisdom to seek care when you need it and to follow it unless you have very good scientific or personal health reasons not to.

Exercise That Works for You

Sitting at a computer all day is not so great for the physical body. We programmers have less risk of injury from moving heavy loads than many other professions, but human bodies did evolve to need some activity.

One of the unexpected side effects of the space program was the discovery that the muscles of humans who don't do the everyday work of moving around at the bottom of a 1g gravity well atrophy very quickly.

Sitting in a chair all day isn't as bad as floating in space, but it's still a recipe for atrophy.

Which exercise to do is a combination of need, capability, and interest. There is no one exercise regimen that works for all.

Every form of exercise has risks of injury. This is the kind of need that evolution tends to solve statistically, not individually.

So yeah, you should move those muscles, but you might tear them or mess up those ligaments.

Or okay, maybe that one exercise worked, but there are a bunch of connected other muscles that aren't doing anything with your current exercises, so maybe they'll complain by causing a pain at some inobvious part of your anatomy and then we're into physical therapy.

Or, this one may be a good exercise for your heart, but be careful with it if your blood pressure is high, and make sure not to do it if the air quality is bad.

Walking is good for you, but if you get bored on a treadmill or strolling through the same routes, maybe you'll just do a little less today, and a little less than that tomorrow, and so diminish.

The combination of physical need and mental interest shows up in exercise almost as much as it does in eating, to some extent for the same reason. We eat to supply materials for building. Exercise is part of the process of building and maintaining the body. In both cases the mind needs to enjoy what it's doing or the body won't get the work done.

Find what you can do and enjoy and do that when you can, hopefully with good medical advice overseeing what you are doing.

As for me, it's Tai chi, dancing, and swimming.

Physical Ableism

The medical model of a single perfect way to live a healthy life may have expanded in possibilities but it still exists and it is actively harmful to just about everyone. There are medical and cultural presumptions of what every person should be able to physically do and anyone who can't do those things is looked down on as less than human, sometimes not human enough to keep alive.

This ableism is a serious danger to a great many people. Internalized, it leads to self-shaming and harm and feeling like an imposition on others. Externalized, it leads to denial of medical care, demeaning of people's lives, insults and abuse, second-class citizen status, and threats of euthanasia and eugenics.

Programmers and users with physical disabilities often need specialized hardware and software to make it possible for them to do their work.

These are open and evolving fields. The concept that software should have built-in options for users who need alternative input and output methods is growing throughout the field. APIs for this exist and are in continuous development. But there is a long path ahead to make this a practical reality.

At the current time, social views of human health are in a great deal of flux. The concept of ableism itself has begun to change the language in which this is discussed. Developments in medical hardware and software, as well as changes in physical and legal access, are moving things in good directions.

But there is also backlash and a tendency to disregard the stated and demonstrated needs of the disabled. And there is still a drumbeat of eugenics and pressure toward euthanasia.

The mental core of the backlash is the lingering delusion of a single right model of perfect humanity. This propagated mental error leads to social structures that propagate physical abuse and disregard for shared humanity.

Mental Health and Mental Ableism

The model of single right humanity dovetails distressingly with the social concepts of perfect virtuous behavior, fashionable beauty and omnicompetence, to create the myth of human paragons.

This fetishized view of mental, physical, and social superiority is terrible for everyone exposed to it.

Paragon thinking takes the idea of normalization and instead of using statistical centers as an erroneous model of how humans should look, act, and think, it centers extremes or indeed myths that no human can actually match.

These views are spread by cultural idolization of people who, it is claimed, are fundamentally better than everyone else. They are also spread by stories that have paragon heroes who are either perfectly good or perfectly ruthless, societies don't seem to care which. And since all of the physical, mental, and moral components that make up this myth change with fashions, one generation's paragon is the next generation's villain.

Trying to live up to the paragon harms many people. For art and STEM, the paragon often has two sidekicks that are also harmful to those who do not conform to the central stereotype: the artiste and the mad scientist.

The tropes of the artiste and the mad scientist assume that anyone who makes art or is involved in science will be off in a world of their own, and that they cannot be expected to handle everyday society. But they will produce occasional works of genius that justify keeping them around even though they do not measure up to the paragon.

Anyone who lacks this genius production capability is not given the leeway in behavior that the artiste and the mad scientist are afforded. Those who do not fit in have historically been deemed insane and in need of being kept away from society, either by locking them away or killing them.

This produces a double-edged risk. One has to be eccentric enough to not be forced into being measured against the mythic paragon, but not so eccentric or non-productive as to be shoved away from society completely.

All of this dubious storytelling erases the simple fact that humans are not that simple and that minds are not instances of tropes. All attempts to force people into molds are harmful.

Unravelling these messed-up stereotypes is both a personal and a social need.

The personal need is for each of us to be able to live a mental life that is helpful to oneself as an artist and a human among humans living in the augmented real world.

The social need is more complex.

Both the artiste and the mad scientist are seen as spending all their time in their studio/laboratory so they won't bother other people.

The social pressure to self-alienate can be countered by developing personally helpful work cycles, doing your art when you need to, and not cutting yourself fully out of the rest of the human world.

Art does not thrive without awareness of other humans.

The cliché of the mad scientist is of someone who does experiments without regard for their human consequences.

The cliché of the artiste is of someone making an artwork so obscure that no one can understand it.

These are both exaggerated forms of paying no attention to audience.

Take the time and opportunities to be aware of other people.

There is a false dichotomy between locking oneself in one's work area and going out and doing "normal" things – i.e. trying to fit in with people who are trying to adhere to the paragon trope. That rarely works.

Interact with other people as yourself, not as any of the stereotypes. Don't try to force yourself to do what others assert you should. If you enjoy being with others then find people who share your interests, or who you find interesting as people.

The world is full of humans and art. Awareness of art and audience does not need to be found in socializing, which can be a strain for many people. It can be found in spending time with the works of others. Art can companion.

There does need to be enough human contact with people who see one as a human among humans so that one does not lose the basic error-checking of interacting with people.

Art is made by humans but has no humans in it. Characters in stories and games are not themselves human. They do not have the fundamental chaos and bubbling possibility of human minds that people need to gain the benefits of living among humans.

Even if interacting with humans is a strain, it's needful in order to prevent the mind from falling into the delusion that humans are tropes, or NPCs.

But what about one's mental health relative to the world around?

The difficulty with augmented reality is that one can lose the reality in the augmented.

People who lose themselves in theories and models can spend their time in a world of data and develop a distaste for the complexity of information. This is the same error that leads to conspiracy theories and other such flattenings, erasures, and substitutions.

It is necessary to periodically emerge and look at the world and humanity, unaugmented, to turn the artist's eye upon the complexities of reality and remember that mind is modeling reality, not dwelling in a purer, more perfect world that has no inconvenient information and uncertainty in it.

Reality checks are needful so that we remember what we evolve in and what our art works in.

For some people this is an exercise in humility. For others it's looking at the materials one is working with and asking how they work and what else can be done with them. These are only two reactions out of the billions of possibilities.

All of the above are praxes to deal with isolation and social pressures. They are based on the assumption that what the mind needs can be supplied by the mind or by human interaction.

But each mind arises in an evolved unique substrate of neurology and neurochemistry. That substrate may need help to function in the too-much-information of reality.

The old mental health model proposed that medicine or surgery might be needed to cause someone who did not conform to the paradigm to be able to conform. A great deal of harm was done through that model.

A more recently arising precept was to help people live as themselves with whatever medicine they needed to be able to handle reality and find their way in it.

As of the time of writing these two concepts both exist, and people in need of medicine for mental health are caught in the middle of the conflict between the two.

All the problems of physical ableism exist in mental ableism. A great many people are being harmed by the difficulties of finding doctors and medicines that can help them live their lives.

All aspects of one's health and life need care and attention. It's easy to fall into the art-making habits of dismissing one's own life or thinking that one is immortal. It's easy to see how one thinks and lives as either perfectly normal or haphazardly pathological.

Neither of these is true. You will need to find and craft your own life. With help.

Personal Exercises

1. Examine your eating habits. See what you can do to improve them.
2. Examine your sleeping habits. See what you can do to improve them.
3. Get what medical care you need and can get access to.
4. Examine the social pressures you find yourself under and see what if any praxes or outside help you've found seem to work for you. If you haven't found anything, keep looking and keep asking. Possibilities exist and new possibilities can be created.

Testable Exercises

Everything in this chapter is personal.

19

Working with Others

Collaboration

We touched on working with other people in the discussion of incorporating other arts.

Everything said there is also true if one's programming work is done in a collaborative environment.

However, if you are working as one coder among a group of artists that includes other coders, we will need to dig deeper.

The basic question in this situation is how art can be made as a shared endeavor.

From the dictum perspective the question seems simple. Many works of art – buildings, for example – are the works of many artists.

But from the res perspective, an artwork is a thing dwelling in and evolving in the back of a single unique mind. How can a res be the work of many minds?

There is an art of collaboration where each mind has its own res and each mind's res evolves in response to the ideas and actions of the other contributing artists. If this sounds like the art of discussion, that's because collaboration expands on that art.

Discussion creates first drafts. Collaboration evolves those into later drafts by combining and iterating discussion and artmaking.

Collaboration brings the work together so that the artists together compose the goals and results of multiple artists. At the same time, collaboration collectively separates the work so that each artist is aware of what they need to do relative to the shared work.

The word collaboration means to labor/work together, which often means separation or division of labor.

But how does the work get divided and how is it brought together?

Different arts have different methods.

Collaboration in writing generally involves each person writing part of the text and the other writers going over that writer's parts as an editor, making changes or suggesting changes.

In programming collaboration often involves division of who is creating what data structures and functions or objects within an overarching structure. This process can fractalize into groups, teams, and so on.

This process can be efficient because programmers can work in parallel. But it does require that the software itself be divided up internally. Each coder must leave holes and spaces between the modules they are working on and the modules others are working on and must supply means to access their work in ways that will work with the code others are writing.

DOI: 10.1201/9781003459866-22

This is often done on a dictum level by discussing what functions and data structures one coder or group of coders will need from others.

This is, unfortunately, one of the places where res work is ignored, which can lead to major bugs and inefficiencies.

The assumption in this dictum description is that if the pieces provide a means of doing what is wanted that they will fit together and work together.

But that's not how things work in reality or in humanity. Whenever you put things together, they will always be interacting and composing in ways that are below and above the surface level.

It is wiser to connect and share the res work by the people involved conversing to generate an evolving mutual understanding while they are doing the dictum work of designing and coding.

Human Entanglement

The metaway in which peoples' ways of thinking can come together is more than a bit of a mystery.

Consider two people, each of whom has back-of-mind processes working to generate and evolve their own res. Those people converse, each bringing forth their own interests and points of view, each working from what is obvious to them.

Humans evolved language and art in ways that make it possible for each back of mind to gather in and generate knowledge and understanding from playing with the works of other minds.

That same evolution allows us to be disinterested, bored, annoyed, enraged, distracted or fascinated by the actions of other minds, and somehow developing awareness of the other person and what they are working on.

There's a significant amount of active neurology doing something so basic that there is no usable language for it. But nearly everyone can recognize that it can and does happen, that minds can through the medium of human interaction change how they are thinking about what they are doing into directions that were not seen to exist before the conversation started.

What is clear is that it is not really a matter of minds agreeing with each other or seeing things the same way, but of minds illuminating the unseen for each other, of showing directions and dimensions and sides of things to each other, and over time coming to rely on each other to provide such awareness.

This metaprocess can be seen in the eventual development of communication between people that requires very little data to be passed from one person to another. People who know each other well rarely require elaborate explanations of what the other is perceiving and imagining.

Often one person will quickly sketch out their awareness, while leaving space in the sketch for the other mind. The second mind will transform the sketch to a more shared sketch. The sketches pass back and forth until eventually both minds have a shared idea of the work that neither could have created on their own.

This kind of interaction requires a particular kind of mutual respect. The minds involved must exercise the ability of each to see the other person as a unique human whose point of view is illuminating.

This method doesn't work as a purely one-way process. That would be teaching, not collaborating. This process cannot be forced. It cannot be externally demanded. As with all human relationships, collaboration works if it works and doesn't if it doesn't.

Collaboration is not the same as love or friendship. Some of the most successful artistic relationships happened between people who did not like each other (c.f. Gilbert and Sullivan). And functioning collaboration doesn't necessarily last. Some collaborative groups form for one project, then fall apart like entangled electrons observed, each in their own spin, where a moment before they seemed inseparable in a mutual dance.

Composition and Contrast of Art Styles

Even without the full-on entanglement of awareness that comes from deep collaboration, it is possible for people to work together and benefit from their different approaches, attitudes and ways of making art.

Just as different arts do some things better than other arts, different artists can be defter and more nuanced with particular applications of the art than other artists.

This is more subtle than specialization. Each mind can bring more of how it does what it does to particular usages of its art. The mind using the praxes that work well for it will put more knowledge, understanding, and capability into the things that are most amenable to being worked upon in those ways.

When a group of artists are working together, each will bring their own capabilities into the project. Difficulties can arise because each artist is, in effect, operating in their own regions of space, time, and possibility with their own awareness of the spatial, temporal and need extents of their own work.

This can lead to each artist either taking over space the others need or contracting their art into overly compact regions and losing the benefits of the outreach of their work.

There are instances of this kind of problem hanging in museums around the world. On display are paintings done by workshops of artists, each of whom rendered a single figure or section of background excellently, and yet they do not seem to all be in the same space. The motions and angles and flows do not come together. There is accidental contrast where there needs to be deliberate composition.

The same can be seen in instances of acting, where an actor would be playing their character well if they were alone on stage or if they were playing opposite a different actor than whomever they are interacting with. The term for this in acting is that the actors lack chemistry.

Which brings us back to our bonding awareness of too much information. Artists are human. Their interactions happen in the real world of information. Their work, individual as it is, needs to bond with the work of the others to affect and be affected by them. Even if the artists do not bond, they need to make space in their art for the work of the others.

In programming this can be easier in many other arts because we can share what's going on in our work by sharing our code and explaining what we're doing and why. We can talk through possible compositions so as to make the parts work better together and play off of each other.

There are three metapraxes that can bring minds together on a subject or project even if the artists do not mesh well together.

1. Respect. Be aware of each other's capabilities and understand where and how each person is bringing their mind to bear upon the problems. Ask questions when you do not understand, and ask them of the people who show that they do understand.
2. Listen. Pay attention to what each person involved is saying, what questions they are asking, and what assumptions they are making that may need to be deconstructed.
3. Move slowly through the spaces you understand well and show your work clearly. Illuminate what is obvious to you, which may not be obvious to others.

As long as the process of development and coding has time and space for the artists to be able to discuss and work relative to each other, this can work well.

It is important to understand that this methodology of mutual care and sharing runs counter to the simplistic assembly line division of labor model of factory work that was developed in the industrial revolution and popularized by Adam Smith in The Wealth of Nations.

Assembly line thinking takes away the idea of humans as individual artists and makes them the prototypes of robots, each trying to do one task to exact data specifications.

Assembly lines are optimized for robotics and are bad for humans, both physically and psychologically. Smith himself said that his model did not work for artmaking. Unfortunately, he did not have the view that everything humans do is art so he imagined that some work could be done without regard for which humans were doing it.

Assembly line thinking does not work well for any creative or human endeavor because it confuses specialized human skills with materials and techniques. It also tries to replace personal vision and measurement in a world of too much information with the robotic ability to machine things to match close enough to data specifications.

The assembly line relies on the idea that each part will be interchangeable with every other instance made by the same part. Interchangeability of parts was the goal of this proto-robotic metapraxis.

This can be beneficial when putting together mass-produced objects, especially when real robots are doing the assembly. The art in mass production lies in the design and implementation and the arrangement of the making of each piece. The artists are the designers and the makers and maintainers of the machinery.

But interchangeability does not work well for humans.

The reason that automation of tasks grew in this economic environment of interchangeability is because such an environment does not need humans to do the work. Indeed, it is dangerous for humans to do such work. The requirements of repeatability do not work well for thinking beings whose bodies can only be so repetitive and whose minds will grow weary if they are forced to do the same thing over and over.

Artists need be able to change what they are doing as they see new things. They need to be able to be inspired to change by the ideas and implementations of the works of those they are working with.

An assembly line can be a lonely place in which people can work next to each other and never come to know each other because they are supposed to be as interchangeable as the parts they are working on.

Any art made this way will have fundamental compositional flaws and a myriad of missed opportunities to make something better that comes from the shared minds working upon it.

Studio and Business Work

Artists working together to make art can be a group of people getting together, enjoying what they're doing and hoping that others will enjoy it as well. They can also be people brought together to make art to sell the art. We'll talk about making money in Section III, Chapter 22 (Foreshadowing: That chapter will not be much fun).

Legally, a group of artists brought together to make art to sell can be a corporation or a partnership or a number of other structures.

If you're going to form such a group, you'll need to consult an appropriate lawyer who knows and practices this kind of law.

We're going to focus on a non-legal, artistic distinction: Is this cooperation a studio or a business?

In this context, a studio is run by people who are skilled in one or more of the arts being practiced and who take active part in the making of the art, whereas a business is run by people who are not skilled in the art and don't take an active part.

A business, from the perspective of the artists it employs, is a single client commissioning works from the artists. A studio is one or more artists employing other artists to help them make works.

It is possible to have an artist running a business and not getting involved in the art. But that's kind of rare. It's really difficult for artists to not bring their knowledge and understanding on work that is supposedly coming from them. They may not pick up a brush or chisel or type a line of code, but they'll likely be nudging everyone else.

This type of business will mostly be covered in Section III, Chapter 20, in the discussions of clients.

We're going to focus here on studios because this book is written for people who see themselves as artist-coders rather than hands-off business owners.

There is a basic tension present in any studio. The artists running it will have a sense of how they would do things, and will have their own res and favored practices that they may presume all other practitioners of the art will share in.

In short, the artists running a studio may push their ways on the artists working for them.

Running a studio is a process of working with other artists. It involves all the metapraxes and concerns we've already brought up in this chapter.

There are other factors you as studio runner will need to be aware of.

If it's your studio, you are employing the other artists. You have responsibilities beyond working with them artist-to-artist. You will need to do work to create a good work environment. There will be more about this shortly.

Apprenticeship is a remnant from an older school of artist studios wherein experienced artists would take in less experienced artists and teach them the craft and help them develop their praxes while making use of them in augmenting the more experienced artist's work.

This system still exists in many modern businesses. Chefs, for example, often work this way. Arguably, academic research still works this way, with grad students and postdocs apprenticing to professors.

Historically, the apprenticeship system varies between grueling but instructive and just plain exploitative.

The system is less workable in coding than in other arts because our art is still rapidly evolving. Fresh-out-of-school coders are likely to know a fair amount about the most recent developments in the field. They bring both the need to learn and new ideas to share. In our art, at least, the teaching can be omnidirectional and we all benefit from that.

If you are running a coding studio, you should make space and time for the exchange of knowledge, understanding, and capability, new and old mutually teaching and learning.

Studio owners are also at risk of succumbing to auteur theory.

Auteur theory is the idea that every work of art with multiple artists has a single real artist. That one artist has the true res.

Auteur theory has been most often used to claim that directors are the "real" artists of movies (even though their work is purely behind the scenes), that architects are the "real" artists of buildings (even though they do not put together any parts that people really live in), and so on.

The assumption in this theory is that whoever is in charge of the largest-scale aspect of the art is the only artist who matters.

This theory erases all the interaction between artists necessary to make any large-scale art work.

The people operating on the largest scale are going to need to be able to listen and correct their art based on the concerns and awareness of everyone else involved.

The larger the vision, the more materials, techniques and praxes will contribute to the work. The larger the scale, the more entangled the information and the more artists contribute to the work.

Far from having a single unifying vision carved in stone, the artists on the largest scale are the ones whose res need to be most flexible and whose dicta need to be the most changeable. The broader the vision, the more evolution will happen and the more errors will show up that need debugging.

The artists on the largest scale are the ones whose art gains the most benefit from the works of others. Rather than a top–down command structure, the auteur is the one whose work undergoes the most editing and transformation because of the contributions and entanglements of all the other artists involved. All the uniquenesses flow into the auteur's work and makes it the most communal of the arts involved.

The Myth of the Fungible Programmer

The auteur theory and the story of the assembly line together create a concept that all the non-auteur artists working on an art project are themselves interchangeable. The myth is that as long as the person hired has the needed skills and can follow orders, the work produced will be the same as it would be if someone else were hired to do the job.

This idea of fungibility only factually applies to data-based things, not to information-based reality.

Real objects and especially people have too many complexities and variations to ever be fully replaceable.

The more skill people have in an art, the more aware they can become of variations in materials, techniques, tools, and especially other artists.

The more we learn the less real fungibility seems.

Audience members know this. They know which artists and groups of artists they like. They know that those artists can't just be replaced. Bands can change radically when one performer is replaced. Plays change a great deal at the replacement of one actor. The more an audience enjoys something the less they will accept substitutes.

There is the concept of a backup singer or an extra whose work can be swapped out for someone else, but even that is not so simple. Differences flow from changes in people. Even something as theoretically fungible as assembly line work can be changed by one person working at a different rhythm than the person they are replacing.

Fungibility of humans is an illusion perpetuated by trying to make people fit into artificial roles and blaming them for not meeting the expectations of acting like the last person in the job.

What usually happens is that the new person tries to conform to expectations while inevitably changing what is happening around them to meet their ways. Everyone else involved acclimates to the new arrangements and smooths over the change in the way they remember things. How much things change over time is hard for the people involved to measure. The old normal becomes the new normal and the minds gloss over that it's not the same normal.

It is audiences who can most tell the difference, especially if they go back and look at or listen to earlier work and compare how things are now to the way they were.

The myth of fungibility needs to be erased so that when contributing artists change, those remaining can be aware that how things were done is not how they will be done. Everyone will need to work to create a new way that works for everyone. Together they can recalibrate to new possibilities and methods and produce new works that their audiences will know are different but still might interest them.

The Semi-Myth of the House Style

If we remove the concept of fungibility there is still one remaining factor that produces continuity when artists are changed, and that is the ongoing conversation and shared praxes that produce what is often called a house style.

Every studio develops ways it does its work together and a shared eye for how things should look and work.

The house style becomes a guiding process, something to work relative to, a sort of fuzzy measurement of works the way we work, sounds the way we sound, tastes the way we taste – or the way what we make tastes, anyway.

House styles clothe the discussions and the developments of res and dicta. They exist in people talking with each other. They evolve over time as people work together and as projects are made and received by audiences. They form part of what is called institutional memory or oral tradition.

House styles are sometimes accompanied by prescriptive manuals delineating what one is supposed to do in what circumstances. This is an attempt to preserve the style even if there is a lot of turnover in the participating artists.

This can work if there is enough flexibility of usage in these manuals. Such manuals can be treated as ways of talking about the art, and this can be effective.

Giving a language to what the artists are trying to do can make it easier for new artists to get a sense of what region of possibility the studio is working in. This can benefit the necessary in-house conversations about whether something is working or not, or give guidance for which ways to go when people are lost.

House styles evolve over time as the artists fit together in how they are doing things and develop a history of what they've done. They are expressions of a developing culture or subculture.

They are not confined to art studios and businesses.

Cultures develop house styles in clothing, language, and other ways people live. Archeologists look at how given peoples built their homes and organized their settlements. These are also house styles.

And yes, that both is and is not a pun.

Each artist who comes into a space that has a house style will need to do a combination of fitting their way of working to the style and evolving the style to work with the way they work.

This is not always possible. Each artist, each human can only work in a space of styles conformable to their basic ways of working. If they cannot adopt the style nor the style expand to fit their ways, then they will not be able to work in that space, no matter how much they might like to.

And that needs to be okay. Human uniqueness and cultural development create niches that only some people will fit in.

We need to accept our own non-fungibility and realize that no one can work with everyone, nor should anyone be forced to try.

Good and Toxic Work Environments

The quality of an artist's work environment can be measured in how well it makes space for the people working in it, how well it transforms relative to the people that enter it and how well it helps them fit the way they work into it.

This applies to a person's individual work environment, not just to shared spaces. It is entirely possible for someone to make no allowances for their own needs and fashion a space that is inimical to them because they think they should be able to work in a set of conditions that are "normal."

The four broad aspects to environment are space, time, possibility, and interactivity.

Space is about the ability of a person to get where they need to go and move as they need to move. It is about the presence or absence of environmental conditions such as air quality, airborne infection risk, and noise level.

Space is a factor in dress codes. Clothing is about motion. How people can dress affects how they can move. Clothing can also be a matter of personal identity, and restrictions on those can enforce social stigma.

Time is about when people need to be where, how strictly they must account for their time and what schedules are treated as ordinary versus extraordinary. A common example is that businesses will often create a calendar that assumes that people will want time off for majority cultural and religious holidays (see Scrooge and Cratchett) but treat minority religious holidays (or days of rest) as extraordinary intrusions into the rhythm of the business.

Time matters for each person's sleep and work cycles. Making it possible for people to work the hours when they are most functional instead of adhering to an old factory shift model of timing can be a great improvement in people's lives.

Possibility is about what ways the artist and the business/studio can benefit each other, what opportunities and aids the business provides and what use can the artist make of those opportunities that helps the business get the art it needs.

Possibility is often pared down by businesses seeing employees as liabilities because their salaries and other benefits show up as minuses on balance sheets whereas the work they do is not seen as connected to whatever profits the business makes. We'll talk about this a bit more in Section III, Chapter 22.

Possibility is often limited for some workers and expanded for others based on unexamined biases. The people running the business need to be aware that no matter how well intentioned they are, they will have and will act upon unexamined biases.

This should come as no surprise to programmers. Intent does not solve problems or prevent bugs. Bug reporting and debugging need to be part of any work environment. Therein lies the basic need for *interactivity* between each person and the other people within the workspace.

What processes exist in the environment for people to talk with and listen to each other? What does the business or studio do to encourage or discourage discussion and human interaction?

The simplest form of silencing is not listening. If people discover that no one in their workspace listens to their ideas or concerns except perhaps to swipe credit for them, they will quickly give up on the thought that their art and ways and lives actually matter to anyone around them. The business or studio will lose an artist.

Perhaps the person will go elsewhere or perhaps the environment will have transformed the person into someone who does as little as possible.

There is a story that businesses care about the bottom line, but that is demonstrably false. Most of the time a business environment is created by who and what the people running the business approve and disapprove of. Who is treated a human and who is seen as less than human? What kinds of behavior are seen as abusive and what is just something people are supposed to go along with? Who gets credit? Who gets blame?

These factors are rarely noticed by the people enacting them or the people who are well-treated by them.

Bias selects for bias. It does not select for good art or competent artists both of which would benefit the bottom line.

To remove the bugs of ingrained bias, the workplace must have processes to help people who are finding the environment inhospitable.

What systems and what people are in place to handle debugging of the work environment itself? Are those systems actually there to handle bugs, or to declare that a toxic culture is really a feature?

A business or studio needs to be aware individually and collectively of space, time, possibility, and interactivity. It needs to have not just an overarching view, but the ability to see it from the perspective of each person there. To do this, each person must be capable of making their viewpoint visible.

The work environment is a collective artwork. Each person in it needs to be both artist and audience for this fundamental infrastructure that determines how well the business/studio will make the art it is there to create.

20

Clients and Audiences

After discussing artists interacting with other artists, we need to shift our attention to two groups of people who are positioned on either side of artmaking: Clients and audiences.

Clients are the people for whom art is made. They appear before the art is made, and the artist must be aware of them during the making.

Audiences are the people who actually make use of the art. They appear after the art is made and the artist must also be aware of them during the making.

This chapter will focus on how artists can interact with these two groups before, during and after the making.

Artmaking can feel lonely and crowded at the same time.

Artist–Audience Interaction

Every person has had the experience of experiencing things in art that the artist did not put in there.

Every artist has the experience of audience members who bring up unexpected perceptions of their works.

Conversations with these audience members can be tricky. Some of the unexpected ideas may be pleasing to hear and some may be distressing. Some may be so enlightening that it will elicit a, "Hey I didn't think of that or at least I wasn't aware that I was thinking of that. Thanks, you've given me a lot to think about."

Some reactions can be fully disheartening, if one realizes that something one made can distress someone because one hadn't thought of something or researched something or considered an alternative possibility.

In most arts, one needs to accept all of this and move on to later works with the benefit of gathered reactions.

In programming we can actually fix problems our audiences have pointed out.

We are that rarity among human endeavors, an artform that has bug reports and updates as part of the process. Changes based on user responses are part and parcel of our work. We can actually correct problems and add or amend features as we go along.

We have the most long-term interactive artist–audience process that exists. We can change the persistence of our artworks of persistence.

And we should be relieved we can. There are few feelings as twitchy as the awareness that one has put something not great in a work one made a couple of decades ago that cannot be fixed (the author says, considering books he wrote a couple of decades ago).

Languages are replete with ways to talk about how things do not work. Each dialect of English has a plethora of words to complain about things. Social networks are overloaded with verbose grumpiness. Product reviews will be explicit about how things went wrong, but terse with expressions of happiness.

DOI: 10.1201/9781003459866-23

This is an unfortunate side to artist–audience discussions. Complaints are easily focused and articulated and a problem for one user is likely to be a problem for many, but enjoyment is complex, spread out, inchoate and unique.

Someone's happiness with your work is not as easy to express as the problems they are having. Praise is found in applause and general thanks, which does not articulate what worked. Happiness has no simple feedback.

There's a good reason for this. What works in an artwork connects to the audience and bonds with their thinking. The rhythms of the art work with theirs. Its interface synchs with their thinking and the results they get make them happy in ways both deep and general.

Happiness elevates and opens minds. Its results and consequences flow through backs of minds and create new inspiration for the art the audience members make and appreciate. The more a person benefits, the less they will have language to talk about the good the work did them.

An argument can be made that two of the pressures toward the evolution of language are the need to assert need and to render bug reports. As a result language is an excellent medium for delivering concerns about art.

But it's not so good for talking about the benefits one receives from the art. Gratitude and inspiration are hard to put into words. They often emerge in a semi-articulate, "Thank you, your work gave me a lot to think about." Or, "That's great. I'm going to use that." Or, "You made me so happy, I can't put it into words."

The depth of meaning is all going on in the back of mind. The conversation is focusing on what doesn't work rather than what does.

Accepting praise means accepting that something personal and unique might be going on in the mind of the person thanking you for your work. The depth of their reaction would take an artwork of their making to express. Maybe your work will help them create such artworks.

Art inspires art and inspires people to take up art.

Distress and grumpiness plod out step-by-step, showing where things are not working.

Gratitude dances and flows and moves too fast and gracefully for the eye to follow. It can be seen in how people make use of the work we have made. Gratitude is vast and deep and requires art to express.

You may hear more bug reports than you hear gratitude, but that does not mean people are unhappy with your work, it means that those who are grateful are making use of what you gave them.

Client–Artist Interaction

Clients are the people who provide the impetus to make art, whether by need or commission.

We presented some clienting elements in Section II, Chapter 15. Then we were looking at it from the clients' perspective where you as programmer were being a client for other artists. Now we're flipping this around. What considerations are there if someone is asking you the artist to make something?

The first consideration is whether you have any reason to do this for this person. Is this someone you know or care about? Is this your boss, or a coworker you're sharing a project with? Is this someone who wants to commission you to make something? Is this someone suggesting an idea that might make sense as an adjunct or feature for something you've made?

All of these can be good reasons to consider making the art.

On the other hand, is this someone who presumes your time, effort and mind are theirs for the mining? Is this someone who thinks you should be grateful for their brilliant idea that they just need someone to make real? Is this some con artist who wants to "partner" with you in some way, legitimate or otherwise?

Don't do art for these people.

All of these, both the good and the bad clients, happen a lot to artists.

Think carefully before taking up a project, even if it's your boss telling you what the next project or phase of project is you're going to be assigned to. Think about the idea and the process. Clarify what exactly is needed and make sure it's feasible.

Clients may or may not know the practicalities of what they want. They don't always understand the art and its current state. It's a good idea to take some time to work through what you would need in order to do what they say they need.

Their idea may not be feasible, in which case you may need to explain why what they want may not be possible.

A person who knows nothing about automotive engineering might not understand the difference between:

"Can you make this car green?"

and

"Can you make this car fly?"

Some clients will accept a yes answer to the first question and a no answer to the second.

Some clients will not accept a no answer. This leads to conversations that can be difficult for the artist because it will be necessary to explain technical matters to someone who may well lack the understanding and vocabulary to comprehend them.

Clients need to understand how much time and what materials, tools and costs (including the cost of your thought and work) this would entail. The less the client understands the field, the less likely they are to understand why you need this much time or those tools or that much money. These conversations can also get annoying.

Clients need to supply specific details of their needs so you can craft the res and then the dictum. They will usually have some idea in their minds but that idea will not necessarily convey what you will need.

A good illustration of this shows up when clients commission visual artists to draw portraits of characters for roleplaying games.

The clients (usually the players of those characters) will often respond to the question, "What does the character look like?" with the character's backstory, which will not provide the needed visual information.

This problem recurs with clients who are not experienced with the art you practice.

Most people don't realize and don't have the vocabulary to understand that their ideas of things they would like are res seeking expression, not dicta waiting to exist. The feeling of the pre-existent art in their minds convinces them that the thing is just a step away from reality, when in fact there is a great deal of work and change to come before ideas can reshape reality. The res needs to evolve as the dictum is created, and both of them can only evolve in the mind of the artist seeking to make the art.

In these first conversations you will need to ask questions that focus both on what the client actually needs you to make and what your needs in terms of time, tools, and materials will be.

If the client is amenable to the questions you need to ask, then by the end of the initial discussion each of you should have some idea of the work to be done.

These don't need to be the same idea. You will see the project from one side, they will see it from another. If you and the client can work together, then over time you can come to enough understanding that the final artwork will be good enough for both of you.

If the initial discussion pans out, it's a good idea to work out any needed contractual details of what is expected when, who owns what rights, when payment is to be given, and so on. Artists who do a lot of commission work generally use variations on boilerplate contracts that fit their needs. More in-depth, complex contracts should be looked over or created by lawyers who understand the fields of art and business involved. If the client is your boss, these should be part of your employment contract.

You will then need to do preparatory work to flesh out the concept in a way that the client will be able to see and determine if your vision and their vision work together.

This leads to the next kind of artist-client conversation, showing preliminary work and talking about where the work itself is going.

If you've made a good rapport with the client about the work, then hopefully the mid-work conversations will lead to corrections on your part, so that the dictum is more what they'd like and need, and corrections on their part of how the work is evolving.

These conversations can vary between really wonderful if the client likes what you've done and has suggestions to make it even better, to really bad if the idea in their minds and the work you've done have diverged even further than the initial state of things.

A great many projects break down at this point, to the dissatisfaction of all involved. Sometimes that breakdown leads to a new beginning with a clearer sense of what could work for the underlying needs. Sometimes the work just dissolves, leaving a few shards of ideas and work behind.

Generally, if the mid-work discussions fit together the work can progress to later stages.

If the client is a large business one may need to present each stage to different people in a hierarchy, which can bring in even more ideas of how things should work which can be contradictory with the work already done. This can also happen with a lone client who does not often see things through to completion.

When, and if, the work is done to everyone's approximate satisfaction, the art is handed over to the client, payments are made, and in most arts that would be the end of things.

But this is software. Bug fixes and updates may well be part of one's contractual obligations. You may need to keep in touch with the client in order to keep things operating, or they may have other people who do that, in which case someone else will need to get inside your work and make sense of it.

If the whole project works out, this client may approach you again (or if the client is your boss give you your next assignment), and the cycle starts anew.

If you are the boss of a business or studio, you need to not waste your workers' time and effort at any stage in this project.

If you have a nebulous idea, then talk things through with your artist employees and see if together you all can make it cogent enough to be worth the work.

If you can't, either give up on it, or let it go and see if you or any of your employees will bubble something related up from the backs of your minds.

If you have a cogent idea, present it to your employees and listen to their feedback and suggestions. They may come up with improvements or concerns. They may point out that what you thought was "Paint it green" is really "Make it fly", or sometimes, that "Make it fly" is really as simple as "Paint it green."

Client–Audience Interaction

In theory interactions between client and audience are not your concern as the artist. In many arts, the client commissions the art, then shows it people by displaying it or singing it or selling copies of it or however your dictum is expressed.

If your client doesn't really understand how your art works or what it actually does, you may need to explain it to the client in language they can use with users.

You may also need to deal with the problems of a client misunderstanding and making claims about what the software does that are not completely accurate. You can find yourself as the artist in the middle between clients and audiences.

It's a good idea to have pre-worked out explanations for what the work actually does as well as some elaborations available for particular features. You may need to have several of these with different degrees of elaboration, depending on the technical knowledge of the audience.

If the client and the audience seem to be getting along fine about your work, you can listen to what they are saying to each other and get a sense of what you have done that is working and how it's working. If invited in, you can take part, but if not let the conversation evolve as they evolve it.

Reliability and Trustworthiness

There are two distinct qualities that are important in client–artist relationships. An artist does not need to have both, but it's good to have at least one and make it clear to all potential clients which you have.

Reliability is the ability to produce work that is recognizably similar to one's earlier works (but hopefully improved) so that clients know to bring you projects and problems that are likely solvable the way you've solved earlier problems. Audiences will know what to expect from your work.

Trustworthiness is the ability to put your whole mind to a particular problem and produce art that works well for the needs of client and audience even if the art is unlike your earlier work.

Clients can bring trustworthy artists those projects and problems that need mindful creativity and innovative methods that fit the particular difficulties of the project or problem. Audiences for trustworthy artists know that the artist may surprise them with each new work but once they get the hang of this particular artwork it will do the job it's supposed to.

Clients can also have either or both of these two qualities or they can have neither.

Reliable clients bring artists similar projects time and again. You know what they'll be asking you for and you'll know whether or not you can do it.

Trustworthy clients have a good eye for new and interesting problems and projects that may require experimental solutions. You never know what they'll bring you, but it's likely to be worth thinking about.

Reliable and trustworthy clients have a care for the needs and abilities of the artists they interact with.

Clients that are neither trustworthy nor reliable are not good clients.

Both artists and clients have a mutual responsibility to be either or both of reliable or trustworthy.

Audiences don't need to be either. Some audiences are reliable in that they seek art in the same space of artworks. Some are trustworthy in that they will give things a try and consider whether they like them or not. But for audiences both of these are matters of taste, not responsibility.

Artists have no call upon their audiences. An audience can be consistent or flexible in tastes and attitudes. Audiences are governed by recurrent personal need and evolving unique taste. Those measurements are perpendicular to the consistencies of reliability and trustworthiness.

Is There a Way of Finding Audiences and Clients?

There is no predictable reliable method to find either audiences or clients for one's work. They inhabit a space of free-floating human possibilities that resembles primordial soup.

There are spaces for showing or selling work. These can help artists find audiences and clients.

But it's all humans and artworks bouncing around, possibly interacting.

There are many people and services that promise to find clients or audiences, but none of them can reliably do so.

There are businesses that offer a selection of art through venues that they supply. Some art does very well from these businesses and venues. Some does not.

Recently, social networks have been used to widen the space of possible connections. These have opened up the spaces of possibilities somewhat.

But like every other method they have also shown only a tenuous ability to find and match artists with clients and audiences.

Methods of opening up passages and environments in which people can search and find possibilities improve the ability to perceive what might be of interest, but do not solve the actual problem. Algorithms that try to connect people do not do much better.

Art is shared through human-to-human connections. That has always been so and will likely always be so.

This is both useful and frustrating. This is a problem that looks like it should be solvable, but it may not be.

This may be another case of a probabilistic problem where the odds can only be played with so much.

It may be that the problem needs to be approached from another direction, that casting it as a problem of finding or matchmaking is an unworkable approach.

It may be necessary to consider unexplored aspects of the problem and work from those.

What those other aspects are is unclear, but it seems to me time to try to find them, because the spaces of artists, audiences, and clients are not fitting together as it seems like they should.

21

Support Network

Home Life. Space to Work and Think

So far we've been focusing on the life of the artists relative to their art and the people who interact with that art. But, contrary to the mythology of the romantic poets who imagined a lonely life in the wilderness or locked away in garrets, artists don't exist in isolation.

Artists can benefit greatly from support networks of people who are aware of their art as an important part of their life and who help them live their lives as artists and are helped in turn by the artists. Mutual care is generally better for all involved.

Support networks vary from artist to artist. Not everyone involved in an artist's life is supportive. Some people close to the artist may object to their artistic pursuit. Others may suggest that an artist's abilities are best applied in a socially normalized structure, such as a business environment, whether or not that works for the particular person.

The presence of people in one's life who are not supportive is a difficult matter. One's ability to live with or interact with such people is a matter of one's own mental health and possibly also physical safety depending on the circumstances.

We're going to look at what and who can be involved in having and being part of a support network.

Let's begin with home life.

Where and how people live and with whom varies a great deal depending on social circumstances, age, relationships, and so on.

What artists primarily need from the home itself is space and time to work and think. Homes hopefully provide a great deal more than that, but the minimum need for an artist is to be able to go somewhere and be able to think and perhaps make notes or do some coding.

Even if one does not work at home, one should not have to leave home to do the res part of one's work. Space and time for thinking need to be part of home life, otherwise that life becomes intolerable.

It can help if the artist is able to do some shaping of the space they live in, to have room for particular memoria, or toys or books or games, room to play, stretch, and rest. All of these can make a home helpful.

For people whose physical homes, assuming they have them, do not provide these, it can be needful to find a home from home, perhaps a friend's home or a café or a park or museum or library, some place where thoughts can stretch out and ideas coil in. It can be hard to make art work if there is no place where activity can flow around one and not necessarily drag one along.

DOI: 10.1201/9781003459866-24

Most people's minds need to be able to lay out their mental furniture as augmentations to the reality they are in. Physical furniture needs to augment the reality of the furniture. If the architecture and furnishings of home life do not provide space for one's ways of thinking and acting, it can be needful to find other places.

Family Support: Illuminating the Artist's Life for Others

Family members who support one's work are wonderful to have, but they may not understand what they are supporting.

The tendency to keep art to ourselves because the cultures wish to treat it as mysterious or alien can and should be undermined by making the processes more visible.

Sometimes this involves saying things that sound strange, such as:

I'm examining this cat skull until it turns into s-curves.

I'm listening to the fire in the stove until I can figure out the music of it.

I'm watching the pattern of raindrops and trying to model the way they're falling.

Artmaking can be illuminated by showing how one is taking in reality, working with it and producing new ways to work with it. The alienation can be removed if one can show that one is looking at and living in the same world as others, though with a different augmentation. Showing how your augmented reality makes artmaking possible can help you share the world you're living in with other people you live with.

Sharing this can also help others perceive their own artistic awareness and help them share how their arts help them see the world. This can lead to the mutual benefits of different artistic perspectives and conversations between those perspectives.

Many people hide their unique awareness from others and from themselves. They are taught to use flattening language that diminishes reality rather than personal illumination that augments it.

People one is sharing one's life with who bring their own arts and perspectives into the conversations can help create mutual awareness discussions. Households of many arts help create multiple viewpoints from which new possibilities can bubble up in everyday conversation. This interacting diversity can also help people not feel trapped into single socially asserted pathways from which no deviation is allowed.

These perspectives and conversations do not have to be between people of the same generation. The discussions can cross generations, helping to form a household in which children will feel more free to experiment with possibilities, find what they enjoy doing, and hopefully learn to see the world in more than one way. There will be more about art and children later in this chapter.

Older generations may have their own arts and perspectives on the changes in various arts through their lives. They may also enjoy taking up newer interests as new things have become possible over time. Experience with change and long memories of errors made and corrected can be valuable for those starting out and needing to learn how to deal with their own mistakes.

This cross-generational conversation has a risk of receiving and passing on generational trauma. People develop habits and perspectives based on the harm they have experienced, and they can pass that on by their presumptions, attitudes and actions. Cross-generational interaction can both bring and disrupt perspective.

With work, play, and mutual interest it is possible to craft an art-aware, art-positive home. And even if not, one will perhaps have shown the people in one's life how one thinks and what might be needful for someone trying to make the art one is making. Understanding can develop even if acceptance is not offered.

Respect for Space, Time, Work Habits, and Artworks

One of the crucial necessities in a support network is respect, respect for people's needs in space and time, for the ways in which each person works, and for the inviolability of each person's artworks.

A household with artists needs to have a principle that other people's art is not to be messed with. Tools, materials and workspaces may or may not be shareable, but art needs to be left alone.

There is a specific cruelty in messing up someone else's work or throwing it away without asking. This should not be done. If one is concerned about the subject matter or attitudes someone is making their art about, it will not help to get rid of the art, doing that will only increase rifts with the other person. If you are on the receiving end of someone doing this, be aware that you are being harmed.

For coding, it helps if you have your own devices that others are not sharing. This may or may not be feasible for monetary or space reasons.

Whether you do or do not have your own devices, you should at minimum have your own backup storage that is not internet-based. Always keep your own copies of your own code (unless there are workplace security restrictions on doing this).

The ability to keep your own art separate and share only what of it you feel you should with who you feel you should is valuable in cultivating the ability to develop your art.

Respect for people making art includes not looking over people's shoulders and not making suggestions unless they ask for them. People need to work as they work and show their work when they are ready.

Talking about One's Art. Listening to Others Talk about Theirs

Being able to talk about one's art with people one is close to can be a help even if you normally have difficulty discussing anything.

It's a subtle skill to develop. One's art usually seems like a mixture of the obvious and what was inspired by others. There is a temptation to discuss the inspiration and ignore the obvious.

But what is obvious to you is likely to be the most potentially interesting and inspiring to others. We gather inspiration from other people's obviousness.

It can take work to explain what is obvious to you, especially if the people you are talking with lack knowledge in the particular area you are working in. Giving background knowledge can be a fair amount of work, but it can also help clarify to yourself what you are doing and how you are seeing it.

It's important to be aware of what the people you are talking to know and how they see things. This can be both easier and more frustrating with family and close friends.

Each art, each field of learning has esoteric knowledge that creates new forms of obviousness.

For example, someone with sufficient mathematical training knows that space can be seen as a three-dimensional coordinate system. Such a person can consider relative positions using such a system. They might look at the world in rectangular, spherical or cylindrical coordinates. Someone without that training would have a hard time with someone talking about positions as three numbers.

Someone trained in the differential geometry used in general relativity would be aware that the Euclidean 3D model just described, while obvious, is also not correct. They are likely to still use that model for short distances, but will have a very different obvious view of the universe.

There are fundamental learning gaps between what is obvious to a person who does not know any analytic geometry and what is obvious to a person who knows analytic geometry, and more gaps between that person and someone who knows differential geometry and general relativity. Each of these people has a different sense of what space obviously is.

There are obviousness gaps about colors between someone who does no visual art, someone who paints as a hobby using pre-mixed paint colors, and someone who paints professionally, mixing and layering their own paints. Each of these people has a different sense of what color obviousness is. Add someone who understands the quantum mechanical aspects of color to the mix and the obviousness of color explodes in different meanings.

If some aspects of obviousness would take too long to explain, it can be wise to focus on other parts that are shorter journeys will lay the groundwork to be able to discuss more later. Give a precis of how you are looking at things. Accept any confusion on the part of those you are talking to and try to clarify. The goal is to lay enough groundwork to be able to explain what you are working on without drowning those you are talking with in technicalities. You can use metaphors and models, just make it clear that you are simplifying things for ease of communication.

If you are on the other side of such a conversation, listening to someone you care about talk about their art and you do not understand what they are bringing up, ask questions. Focus in on what you find interesting in what they are doing and in what they are trying to show is interesting to them. Be guided by the human interaction rather than the technical. You can always look up the technical aspects later and then ask clarifying questions.

It can take a long time to learn about other people's arts, but doing so can give you other perspectives usable in your work as well as the ability to ask for help from someone who understands and uses this perspective.

Art and Children

Children need opportunities for art at home, if possible, and in school certainly. One of the ways that art and STEM are put in artificial conflict is in school curricula. The belief that STEM can be encouraged by removing art is foolish on all levels. A child who enjoys music can play with the generation of sounds on a computer. A child who draws can learn about the digitization of pictures.

If you have children (or are related to children) who are artistically deprived at school, you may be able to supply them with materials and tools at home. Online tutorials are available for a variety of arts. Some arts are difficult to do without specialized equipment, but a fair amount of learning can be done these days through computers. Hopefully, the physical materials and tools will become available to them in time.

Children can be taught artistic awareness and augmented reality quite young by showing how things really look and helping them listen to the world around them. I found that involving children in cooking, if that is an art you do, works very well for many kids.

And, of course, you can show them aspects of coding early on. The ability to cause things to appear on computer screens or sounds to emerge from speakers delights many young people. Letting them get inside these processes (the software certainly, not necessarily the hardware) and see how things work and can be made to work can stir a wide diversity of arts quite early on.

While this book is not focused on teaching coding to children, the perspective of seeing programming as an art can be imparted early by anyone who enjoys the art. Teaching that programming is art can make teaching children easier.

Too many programmers see the work as so serious and needing such care that they think there is no way to impart the skills to children. But art is playful and coding taught as play can fascinate and inspire kids for the rest of their lives.

Social Life, Friends, Your Art, and Their Art

Friends come in many varieties and share different aspects of one's life. With some friends it may be easy to talk about one's work and art. For others less so.

This is one of those things where the particular arts one practices can make it easier or harder to converse.

The arts that are seen as arty have a fascination all their own. A visual artist can usually share their recent works with most of their friends.

Writers can talk about their work, but sharing it is trickier. Giving a book one has written to one's friends can be seen as burdening them. The same applies to other long form arts such as animation, podcasting, and video making.

Singers and musicians used to be sought out socially to provide entertainment for family and friends. This was before everyone had access to recorded music. These days it's less common to ask others to sing or play.

Programmers are in the same boat as writers and video makers. Talking about one's art socially can be seen as an imposition and distraction from other subjects. And asking people to test your software can be seen as imposition.

The main thing is that art should not be forced on people. Long form art takes commitment to interact with. If you have a group of friends interested in your art and you in theirs then it is quite possible to discuss one's work. But for groups of friends without those shared interests you may need to keep your work in the background and share other things with them.

There is a curious paradox brought about by physical distance and separation. Friends you share physical space and time with would need to take time away from what they are doing with you to look at your art. But friends you interact with online who express an interest can be sent a link to try your work at their leisure. The distance of the internet makes it easier for someone to ask and for you to provide your art without the pressure that proximity can generate.

Colleagues and Other Uncertain Supports

Colleagues and coworkers exist in a complex social space. They will usually consist of a mixture of people one likes, some one dislikes, a few real friends, and various acquaintances.

Interactions with such groups often focus around work.

But there is not exactly a support structure in sharing workspace and time. There may not even be any social structure as such. People brought together by a common focus and enforced proximity do not automatically entangle. Working for the same company does not engender caring or trustworthiness or interest in each other's lives.

There are other societal groupings which may or not be actually social and supportive, fellow members of a religious group, sharers in a hobby or fandom, gaming groups and so on.

Sharing does not inherently produce caring. The mixture of people in these groups can disappear from one's life if one gives up the focus around which the group is centered.

This can be very jarring, as it feels as if a support has disappeared from one's life.

Some people continue with such groups even if they have lost the focus because they are afraid of losing the people, but that isn't good for anybody. People who are with you because you belong to the same group are not with you. You and they are with the group.

Learning and Keeping to Your Social Tolerances

To make use of one's own support network and be part of the support network of others requires a level of personal awareness and care. Social interactions are praxes, and too much or too little of particular kinds of interaction can have detrimental effects.

People need to learn their social tolerances. It's necessary to develop a sense for which ranges and densities of human interactions energize them and which drain them.

In theory, one should be able to try various activities, see whether one likes them, home in on the aspects one enjoys, and mostly do those. But there are difficulties.

Companioning the lonely is a complex and individual need that societies often presume has clear, set, obvious forms that all should follow. For example, many societies have socially approved fandoms (sports are common) and socially disapproved ones (roleplaying games used to be such, but that's changed).

Fandoms are all structurally the same, but the activities that fans follow are individual to each group.

Two people may have opposite reactions to two fandoms, each enjoying their time in their favored fandom and disliking the activities in the other. There's a clear symmetry. I like A, but dislike B. You dislike A, but like B. Okay, that should be easy to socially accept and plan accordingly.

This symmetry is broken if one of the fandoms is socially approved and the other socially disapproved. Participants in the disapproved fandom may find themselves pressured by society to leave it and join the approved fandom, whether they would like it or not.

There are a variety of social pressures about social possibilities. Finding the possibilities that work for you is made trickier because going along with a social pressure can feel like a relief and going against one may feel draining.

Essentially, social pressures bias the measurement of personal enjoyment in social situations. Embarrassment can mask one's happiness if one is constantly worried about being seen doing something socially disapproved.

There is also the effect of normalization and alienation. A person can learn to think that their misery in doing a socially approved action is fine and their enjoyment of a disapproved action is shameful.

This is a situation where help from the supportive parts of one's support network can be highly advantageous. People who know one well can be pretty good at telling whether one really enjoys something or one is actually unhappy but trying to keep up appearances.

It is also possible to seek outside assistance if one's whole social network is immersed in the same social pressures that may be distressing one.

This is important. Finding out one's actual social tolerances is essential to crafting a life that one enjoys living, in which one can make the art one needs to and likes to make.

22

Making Money

Nonmonetary Systems

Everything humans do is art.

Art made can be given to other humans to help them live their lives.

But how does the artist live? How, if one is making art, is one's hunger fed, one's nakedness clothed, one's homelessness sheltered?

A problem this old has been solved many times with recurring methods.

The most basic method is communal. People in a community make art that helps everyone in the community. The food gatherers and cooks feed everyone. The tanners, weavers, and sewists clothe everyone, and so on.

This works well on the small scale. It relies on everyone knowing everyone else, having a sense of individual and group needs.

The communal solution breaks down when there are too many people involved. The necessary information on need and satisfaction of need is not generated or learned.

Communal art is still the most common way art is created and distributed.

But communal art is not paid much attention to when people talk about art.

Conversation, home cooking, telling jokes, acts of love and friendship, all of these are communal art, but they are rendered invisible in the economic world view (Sexism is also involved in the invisibility).

Individual artworks can be made as gifts between friends and loved ones. This works in a more intimate space than the communal and usually involves deep entanglements of mutual understanding between the people involved.

This kind of art is romanticized and often dismissed from artistic analysis with the phrase, "That's sweet."

Communal and intimate art rely on a fundamental symmetry of artists and audience knowing each other well.

This symmetry breaks when the artmakers do not know the needs of who they are making their works for. Artists in this situation make art for an abstract imaginary audience that can somehow make use of the art.

Such art can be traded with other people who also make art for abstract audiences. This trade is usually art for art, the barter system.

Barter still exists and is more common than people think. Trading object A for object B is the simplest form of barter.

DOI: 10.1201/9781003459866-25

Barter moves into a more abstract space owing to the human ability to imagine the future. This transforms the trading of objects into the trading of favors. "Do this for me and I'll owe you one."

Barter is still a person-to-person action, whether it is bartering in present or future. It does not scale up easily.

Because most of us grow up in money-using cultures and money can occlude awareness, it can be a useful praxis to observe how much communal art, gifting art, and bartered art is actually going on in your life.

Now let's look at money.

Money as Tool and Problem

The general economic story of the development of money is the concept of the medium of exchange. The idea is that barter gets complicated if I have A, you have B, and them over there has C. I need C, but them over there wants B, and you need A. So I have to barter A to you for B and then barter B to them for C.

Money is a tool to simplify this. Each of us buys and sells for a fungible medium of exchange, so we don't need to do complicated round-robin transactions which can break down if our needs aren't coordinated in time.

That's money as tool. It's a brilliant, recurrent, useful abstraction.

It also has a lot of problems.

The first difficult aspect of money is its simplicity.

Money is a single datum used as a medium of exchange for all the complex information of human life. Money creates the illusion that everything should have an abstracted single-dimensional value which we all should treat as an objective fact.

This myth is backed up with the heavy mythology of economics.

I'm not going to get into a major rant, tempting though it is. I will bring up one concept in economics that relates to our earlier discussions of information theory.

There is an idea called the efficient market hypothesis, which says that every object has a single appropriate price based on current conditions of supply and demand.

To back up this idea, classical economics models each person as a completely rational decision maker with access to perfect information about all economic situations.

Both parts of this are clearly absurd.

The first part, that each person acts as a completely rational decision maker, cannot be given a meaningful definition, especially given human uniqueness and personal taste.

The second part, that each person has access to perfect information about all economic situations, is a violation of the laws of physics. It is not possible to know all relevant information thanks to uncertainty and entanglement. And as we've seen, relevant unfindable information grows with complexity, so that by the time one reaches the human scale we are awash in a sea of too much information for us to process.

However, disproving a theoretical justification of a system does not remove the system. Money is a tool common in all modern societies and we are, for now, having to work with it.

So let's get as practical as we can in a world where people are paying attention to a single datum and ignoring the complex information they are part of.

There Is Not Now nor Has There Ever Been a Single Way to Make a Living with Art

The phrase "to make a living" is pretty horrific if you think about it. It puts the tool of money as more important than the lives of people. We're making a note of that and sailing around it, because vast mountain ranges of books on morality and economics and other distressing topics.

We're going to stick with the small scale practical question of one artist making money while also being aware of how bad the underlying structure is.

There are recurrent methods of making a living with art that artists throughout history have employed in order to live and enjoy themselves.

Options for how one goes about this are dependent on social structures and possibilities. At the moment, the systems are in a great deal of flux and nearly every method is available in one way or another.

We're going to outline the possibilities but not get into details because there are too many variations possible for each. The goal here is to give you a two-steps-back perspective on the possibilities so that you can consider what might work for you in your personal context. Using that, we hope you will be able to take a step forward and try what might work for you.

Commission Work

The essence of a commission is that a client pays the artist to have one or more artworks made for them either to use personally or to market.

There are legal issues about rights to reproduce and sell, some of which we'll talk about in Section III, Chapter 23. Anyone doing commission work should educate themselves on these matters.

Commission work is very erratic. You never know who might be interested in and capable of paying for your work. Indeed, there is the problem of potential clients knowing your work exists.

It's also hard to know whether one commission will lead to others or not. Commissions can be a high-stress existence for an artist if it's their main source of income because it's very inconsistent. Doing commissions as a sideline can work better, but that's also complicated. Commission work is irregular and when it shows up can consume a great deal of time and energy.

An attraction of commission work is that a lot of clients have interesting needs that can't be met by shopping around. The work they would like is often more personal or niche than what can be found in larger marketplaces. Exploring such niches can really expand the depth of the artist's knowledge. My wife, for example, has done scientific illustration commissions that showed her how chimpanzees use tools.

Seasonal or Part-Time Work

There are fields that have need of artists to do certain kinds of work on a recurrent but not constant basis. People may be hired to do specific artworks for holidays or during the summer or for the run of a play. Basically, this is work where a demand can be anticipated and will last a stretch of time and then stop until the next season comes around.

Part-time work is a variant of seasonal. It may not be as focused on predictable time periods, but the same precepts of working a stretch of time and then leaving the work behind apply.

What seasonal work is needed is something that is not always well known outside each particular field. Seasonal work has the advantage of being longer-lasting than commission work, but is, well, seasonal. The income it generates is only for the stretch of time it works for.

As with commission work, seasonal work gives access to niches of society that are often not well known and gives opportunity to expand one's knowledge and learn new ways one's arts can be applied.

For programming, seasonal work can show up with studios or businesses that need extra hands for projects they are working on that are reaching crunchy periods. This is usually not a good work environment, owing to stress and lack of training time.

Long-Term Employment

Most modern societies were, for a long time, structured around the idea that most people have long-term jobs that pays them consistent wages that they can rely on and plan their lives relative to.

This idea was never really true, but even the veneer of it has been worn away over the last several decades.

In theory, one is hired to do a job by an individual, business, or studio.

The job will involve working on either in-house systems or projects to be made and sold by the company.

In-house systems work involves keeping the business itself operational by creating or maintaining its infrastructure.

Projects to be sold depend on what kind of company one is working for and what products or services it provides.

Long-term employment as an idea was sold on two points, that it provided a reliable source of income and that it provided the chance of promotion to better jobs. This was mythologized into the concept of climbing the career ladder.

There are mountains of analysis deconstructing this myth. In a nutshell, the information says that you need to be wary. Long-term employment advertises itself on security and advancement. But these are selling points, not honest descriptions.

There is a great deal of cultural and corporate mythology and a lot of language created to enforce the myths of long-term employment. You may luck into a good working environment with decent wages and benefits, but it's unwise to think that such a situation is permanent. Long-term can end quite abruptly.

Long-term employment in a good environment can help you develop your art and your awareness of needs. But don't feel defined by the business and what you are doing for it.

Marketplaces

Marketplaces are communal spaces where people offer what they have made for sale.

In theory a marketplace is a social good because it puts various needs in one place where people can go and find what they personally can make use of.

Commercial districts in towns and cities are marketplaces, as are online stores.

Selling one's works in marketplaces is a basic source of income for businesses and individuals and has been for millennia.

The difficulties with marketplaces are largely geographic. How are the marketplaces arranged? Who gets to set up shop where? What are the hours? How do people get into and out of the marketplace? How can one's wares be made visible? How much to charge?

Getting into a marketplace and having your works found and bought are all problems that people confront. These problems have spawned a variety of adjunct industries such as advertising that propose to solve the geography needs for each seller.

Again, there are mountains of analysis about how well these methods work. But there are also mountains of advertising that claims to demonstrate that the methods will work if only your art is worthy. Be wary.

For software, marketplaces went from being narrow spaces to very wide ones. When the only computers were mainframes there were only small markets for software. But those markets consisted of large companies buying software for huge systems. There was a lot of money made on early software. As hardware grew smaller and more powerful, the marketplaces grew and the prices of programs dropped.

It is now not difficult for a lone programmer to write programs and upload them for distribution.

However, the geography problems have grown more complex. The internet is made of data, not information the way physical geography is. Every object on the net is isolated, with only narrow links connecting them and search engines to find them. And internet commerce is dominated by a small set of very large marketplaces. Your work can vanish easily. All emphasis is no emphasis. All advertising is no advertising.

Programmers have also been making use of the nested market-within-a-market of in app purchases. Programs, mostly games, can contain their own products that users can buy to use in the game.

This idea rapidly becomes morally harmful. Some games are structured so that one cannot advance without buying more products, purchases that are really just gambling. Basically, paying to continue is a model that attracts addicts as its customers and is just plain exploitative.

Software as a service is a related concept where people pay to keep being able to use a program. Essentially, the user gets used to the tool and then has to give money in order to continue to use it. This is modeled on renting which has its own long, unpleasant history.

The software marketplace is at the moment full of metaproblems crying out for both social and software metasolutions. It is not wise to predict how they will be solved and what will happen afterwards. But the problems are well worth considering for exactly that reason.

Patronage

Patronage is a system that is coming back into favor in a new form after a long period of decline.

In classical patronage an artist was supported by a person who would occasionally act as a client, requesting specific works be made. The rest of the time the artist could make such works as they felt the need or inspiration to.

This had benefits and drawbacks. An artist could not afford to make art that would be disliked by the patron and certainly wouldn't make works that might offend them. The patron was therefore always present in the artist's considerations about what they might make.

The primary benefit of patronage was that the patron was supporting the artist because they liked the artist's work in general and were trusting the artist to make more things they would enjoy.

Patrons had to be wealthy and for a long time were also politically powerful. Artists could find themselves embroiled in political complexities and their art was often used for propaganda. Pope Julius II's patronage of Michelangelo is a fascinating case study.

Modern patronage has solved many of these problems with crowd sourcing. Instead of one wealthy patron, there are many contributors. Each patron has some access to the artist and some may be able to commission works depending on the exact setup the artist uses.

The difficulty of making modern patronage work is folded into the problem of finding audiences (or audiences finding the artists they might like to support), as well as overall economic conditions, which will always complicate things.

The system still has bugs in it and is still evolving, but this new form of patronage may well be one of the most useful systems for artists.

Societal Support

Societies and governments have at various times and for various arts supported artists. They act both as clients commissioning public works and long-term employers in various ways.

Many programmers work for all levels of government in just about all nations. That's not what we're talking about here.

Societal support is halfway between old-style and new-style patronage. It gives artists the means to live while they work on their art because the society deems that art worth supporting.

This can have all the difficulties of patronage but with even more politics thrown in. Modern societies may have bureaucracies whose job is to give grants to individual artists and art groups. The art those artists produce has at various times been used as the basis for attacks on those bureaucracies.

Most of these bureaucracies work on the grant system (which is also used in the sciences and other parts of academia), wherein the artist needs to apply for a grant while explaining what they will be doing with the money. Arts grants tend to be small these days and can supplement, but not provide an income.

Costs

Along with the basic human needs of the artist – shelter, clothing, medicine, and food – artists also need materials and tools, space and time to work in.

It is a startling fact that nowadays programming is one of the least expensive arts to practice. Ask a painter how much they spend on paints, brushes, canvases, and framing. Ask a dancer how much shoes cost.

Programmers can do great work on a midrange computer and free or cheap programming environments and not-too-expensive internet access.

We may also need specialized input and output devices depending on what we're working on or our personal needs.

But overall, our costs are low. Of all the things to have happened in the development of computers this is one of the most unlikely bits of fact that was once science fiction. Cheap handheld computers. What will those writers think of next?

All of these costs, the human support costs and the art support costs need to be considered when looking at making money.

What do you need to live?

What do you use to make your art?

How much money must you make to make a go of it?

Costs can be disheartening. They can distract from the reality of what one is doing. The debit lines on a balance sheet can create the illusion that the things one is paying for are bad because they reduce the amount of money one has.

This gets back to the substitution of the single datum of money for the whole breadth of human life and experience. People have looked at negative numbers and decided their lives are not worth living.

If you are trying to budget to see what you need and what you need to do for it, please do not fall into the trap of believing that the money is more real than the human need.

Money is simpler than need. That makes it less real than the fundamental complexity of humanity.

If you start to drown in the figures, take a step back and get a breath of reality so you can return with perspective and an artist's eye on the needs you have and what can be done about them.

The social structure may be making your life harder because it is attached to the idea of money as the be-all and end-all of humanity, but you don't need to accept that.

See the reality of life and see that money is only one augmentation to that reality, and not a particularly good one at that.

Money is a useful tool that people have become too attached to. Use it as a tool, but don't accept the attachment and the culture of attachment.

Employing and Commissioning Other Artists

So far we've been looking from the perspective of being the hired or commissioned artist.

But what if you are employing others or commissioning others for individual works?

To quote Rabbi Hillel, "That which is hateful to you, do not do to your fellow!"

In short, don't be the person on the other side of what we've been talking about. Don't be the jerk who undervalues the works of others or who sees people as nothing but negative numbers in balance sheets. Do not ignore their human lives and needs. Do not disregard their input as artists.

If you're having trouble with this, reread Section III, Chapters 20, 21, and this chapter up to just before this section (Do not get caught in an infinite loop by rereading this section and following this instruction again!).

This time read from the perspective of the employer or client. See if temptations to belittle and dehumanize for the sake of ego and profit pop up, and if so, treat those temptations as bugs in the system of human mutual awareness and action.

Having a Studio

The advice of the last section holds even more true if you are running a studio with the added warning, don't be an auteur.

The artists you hire bring their own views and knowledge with them. Don't try to overwrite that awareness.

If they are new to the field and to your studio, help them acclimate and integrate with the others.

If you personally are not good at social interactions, have at least one person on staff who is and can help newbies fit in.

Try not to be the only source of ideas for projects in your studio. Listen to your employees and if they have an idea that sounds good see if you can work it into your work schedule. If so, implement it and give them public credit for coming up with it. If they do the actual art give them credit for making it work.

If it is not possible to fit their idea in, let them know.

Do not discourage them from pursuing the idea on their own. Your employees are not your property.

They may be with you for a time until they can build up the skills to go off and do their own work. That's fine. That's how the apprenticeship system is meant to work.

Give people jobs that they are personally suited for and the opportunity to extend into jobs they would like to try to learn to do.

Be aware that like everyone else you have biases. You may unconsciously be undervaluing the work of some of your employees for reasons wholly unrelated to their actual work.

Pay attention to complaints and concerns. There are likely to be parts of the work environment that are not good for all of your employees.

Never trust the idea that you are doing things perfectly, and never trust good intentions as a replacement for functional methods.

Do not assume that you know what's best for everyone.

Artist Organizations: Guilds and Unions

Economic structures only work because people go along with them. They stop working when people don't act according to the ways they lay out.

Exploitative structures rely on people being isolated. Exploitation relies on each person thinking they are alone or helpless against the vastness of the structure.

That vastness is an illusion. There may be social force used to maintain the structure, but that doesn't make it real or large. Enforcement can be and often is applied on behalf of illusions.

People communicating with each other can discover that they are being exploited structurally rather than being treated individually. People organizing together can work against exploitation.

The history of guilds and unions is a long and complicated one with way too much blood and pain to it. Because of this people have argued against artists forming such groups. But artists can benefit by getting together to share ideas, support each other and organize mass action if needed.

Labor laws and policies vary from nation to nation and region to region, as do what labor organizations exist. Note that labor organizations are not the same thing as labor parties, which are political parties first and may or not provide any actual benefit to people who work.

It's a good idea to investigate what guilds and unions exist for one's particular field of art and work and find out their attitudes and history. There may be one or more that you will find it worthwhile to join. Even if you don't join, looking into the history can show you what issues have cropped up over the years for people working in your field.

Personal Exercises

1. Observe the communal and gifting art in your life.
2. Observe the barter, including trading of favors, going on in your life.

Testable Exercises

1. Document some of the results of personal exercises 1 and 2. Do not document anything you feel is too personal to share.

23

Intellectual Property

Software Is Complicated

I am not a lawyer!

This chapter is what intellectual property looks like from the perspective of someone who has written a fair number of books and quite a bit of code and has several software patents. This is not legal advice. It is meant to give a sense of what dealing with intellectual property is like from the client side.

First, a bit of history.

Intellectual property arose as a concept because there was an idea in Enlightenment Europe that arts, sciences, and inventions should be encouraged because they were overall good for societies.

The method governments chose to encourage creativity was to say that the creators of artworks and inventions would be the owners and determiners of what could be done with those works for a time, after which the works would become common property, something anyone could use.

This concept manifests differently in the laws of various nations and became the subject of international treaties.

I'm not going to try to cover anywhere near all of this. Most of my experience is with late twentieth and early 21st century US copyrights and patents.

We'll also touch briefly on trade secrets and trademarks, which are their own thing.

Software is weird in terms of these laws, which is not surprising considering how old the laws are relative to the recent vintage of our art.

Software is written, and therefore the code itself falls under copyright law. But it is also an embodiment of methods of doing things, therefore the programs, as tools, fall under patent law.

So, when coding you need some awareness of both of these.

Software as Copyrightable

The basic principle of copyright is that if you make a work of art it is presumed to be your work and is automatically owned by you unless proven to be someone else's.

DOI: 10.1201/9781003459866-26

Copyrights have for a long time extended to x years + a life in being, which means if you made it you own it for your life and your estate owns it for x years after you die. The value of x has been changed by law multiple times.

Copyright is given for the dictum of a work, not the res. You cannot copyright the underlying ideas that the back of your mind crafted before the front of your mind crafted anything.

Copyright is automatic. Works, once made, can be registered with governmental copyright offices, which can be necessary if you need to prove that you made it and when you made it.

The ability to sue for infringement of copyright depends on the governmental body that regulates copyright. Even though copyright is automatic for everything you create, in the United States you must also have previously registered your copyright with the Copyright Office in order for you to go to court over infringement. Other nations generally are more generous about allowing creators to sue.

Fees for United States registration are pretty low, generally in the same price range as vehicle registration fees; although this can still be a burden for anyone who has a significant number of artworks.

Copyright can be sold or assigned to another person.

This can be done before the work is made by having a work-for-hire contract, wherein the work you are doing for someone is deemed to be their work. Many employment contracts specify that work their employees make for them is work-for-hire. Certain commissions are work-for-hire, but by no means all.

The right to make copies can also be given by contract without giving away the copyright. The book you are reading is copyright to me as an individual, while the publisher has the right to make and sell copies, for which I get a cut. The term for my cut is "royalties", which sounds a lot posher than it is.

If you are reading a copy of this book published more than x years after I die (for whatever value of x is relevant), then whoever made that copy will have had a right to do so because this book will then be in the commonwealth of literature. For a given value of literature.

The standards for infringement of copyright start out simple enough. No one gets to copy or distribute your work without your permission. No one is supposed to publish work that is derivative of yours without your permission. How far "derivative" as a concept extends is a matter of a lot of messy case law.

There is also the concept of "fair use", which basically says that small portions of a text can be used as examples in other books or for the purposes of publishing critiques and other more arcane uses. Fair use is a legal defense and usually needs to be argued in court, which we'll get to in a minute.

There is also the whole complex field of fan fiction and the idea of derived works that are the same as the original with the serial numbers filed off. This is a volatile field at the moment, so let's just say stuff is happening and attitudes are evolving.

On matters of code, swiping bits of other people's work is not good, but it's become common practice to post snippets of one's own code to show people how to solve annoying little problems and give permission for people to use it.

Anything you own the copyright to you can give general permission to other people to publish, unless you are under contract to give someone an exclusive right to publish. I cannot at the moment of publication give you, the reader, permission to copy this book, because my publisher has exclusive publication rights. Also, I don't want to.

It's a good idea to mark the files of your code with a copyright symbol ©, the name of the copyright holder, and the year the code was written. This is not a legal necessity, but it shows that you knew what you were doing and who owns the code.

Copyright is free. Registration is cheap. Infringement cases are neither. They are legal proceedings and can be really expensive. This is a fundamental asymmetry in the application of the law. Businesses have an easier time defending and challenging copyrights than most people do because of the legal costs.

An individual artist often can't afford to go to court to protect their copyrights. If you find yourself in this situation you may need to give a great deal of thought to whether or not to pursue or defend claims.

Software Methods as Patentable

Patents operate in many ways opposite to copyright.

Copyright is about art. Patents are about innovation.

A patent is in theory given to someone who has created a new way of doing something or makes possible a thing that was not possible before.

In practical terms, patents need patent lawyers and lawyers cost lawyer money. Patents also need to be filed, which costs legal filing money.

I've got a very good patent lawyer. He helps me take what I'm doing, find the parts of it that can be patented, put them into appropriate form and file them. That costs thousands of dollars.

And that's just the beginning. A filed patent application is assigned to a patent examiner. When I started working on patents, I was told by someone with more experience than I had that the job of the patent examiner is to say "No."

Patent examiners are supposed to find legal reasons to deny the patent. The examiner will look through prior art (previous patents and filings) to see if what one is filing is already possible. If so, the examiner will issue a rejection in the patent dialect of legalese, which one needs a patent lawyer to interpret.

One can then give up or adjust the patent claims to overcome the examiner's objections. This is an iterative process that either leads to abandoning the patent or it being accepted. If accepted, there are more fees to pay.

If accepted, a patent is granted for, at the moment, twenty years. But patents have to be maintained. You have to pay maintenance fees on a regular basis to maintain the patent, otherwise your invention goes into the public domain.

The scale of all of this is in the thousands of dollars.

And all of that is the relatively simple part of patent law.

Here's where it gets complicated. To qualify for a patent, the invention needs to be described in such a way that a person skilled in the art (usually an engineer of an appropriate kind) could follow the instructions within the patent and build a copy.

A patent is a how-to application. What you invented has to be makable and remarkable by others, otherwise the Patent Office won't patent it.

You can't patent an idea. You have to patent the manifestable method.

Basically, you are patenting a sufficiently advanced res of a work that someone else could make a dictum from, and that dictum must work pretty close to the way your dictum works.

The patent describes the res, so you don't need to publish the code for a software patent. You don't even need pseudocode. But you do need to describe the process in patentese.

Again, you'll need a patent lawyer because your patent will be written in a specific legal dialect with its own vocabulary, grammar, idioms and drawn diagrams.

The patent claims will be more abstract than the physical or software object, but not so abstract as to defy someone who knows what they're doing from making it.

There is a relatively recent addition to patent case law that says that an invention cannot be obvious to one skilled in the art. The thesis here is that you can't buy two things off the shelf in your hardware store, connect them, and say you've invented the connected object.

Your invention cannot be obvious to make before someone reads your patent, but it has to be makable after they read it.

You can't patent what has been published.

There are a lot of jokes based on the idea that a thing can be patented if nobody has patented it before. This is a popular myth, but it's not true. You can't patent a known process (with some very weird biochemical exceptions … kind of).

But you can patent something that has never been applied before to certain uses.

My first patent took some known methods in differential geometry and applied them to animation and animated figure-making. I did this because a friend of mine who was an animator at the time told me what they had to do to make some models, which struck me as ridiculous given the existence of this two-hundred-year-old math. It was obvious to me, but no one else had made this application of this math.

However, had I gone on the internet as it then was and ranted about how ridiculous it was that no one was using this math for this purpose, I likely would not have been able to patent my method because it would have been published. I'm bringing up an example from an old, now-lapsed patent because if I talked about more recent matters, things could get complicated.

Patent holders can sell or license the right to make products based on their inventions in a manner similar to copying artworks, although with its own legalese and contract styles.

Let's get back to the patent examiners.

Patent examiners vary in training, skill and outlook. Albert Einstein was a patent examiner in his younger years. Some people have looked at this as a reason to sneer at Einstein, as if anyone could do that job. But examining whether something should be patentable was and is a skilled job, not an underestimation of his abilities.

Examiners have to deal with patents on all sorts of subjects from all sorts of filers. They need to be able to make sense of and poke holes in the inventions submitted to them.

It takes work and thought to show them that one's invention is a useful, not overly abstract innovation. The process is long and arduous and not cheap. Patent examination is in its own way an art. Most of the time which inventions are patented make sense, but there are weird oddities that make it through the process and some things that look like they should be patentable that don't make it.

And then there are the infringement cases.

For a patent to be infringed, someone must make something that matches the claims in a patent or is equivalent to them and that word equivalent is a whole complicated legal mess.

Infringement cases are a specialty in law. If you are involved in one, you need a good patent lawyer who understands all the patents involved.

Patent infringement does not require that the infringer know the patent existed. It is quite possible to create something that depends on someone else's patented process that you've never heard of.

Suing for infringement can be risky, because it's possible to prove that a patent should not have been issued in the first place. People can lose their patents in suing for infringement if the defense can show that the invention was obvious to one skilled in the art at the time it was filed or that the patented invention itself infringed on someone else's patent.

This legal muddle is why patent trolling exists as an annoying method of making money. Patent trolls will buy up patents and use them to sue people putting out products that might or might not infringe on the patent, hoping to get a settlement by a company not willing to risk years of litigation on a product they have ready to send to market.

Trade Secrets

Patents have not just one, but two opposites. Copyright is dictum to its res. Trade secret is hidden where patent is public.

I have no direct personal experience with trade secret law, so this section is not going to be anywhere near as elaborate as the last two. What I'm about to say is a lay person's distant understanding.

Suppose you invent something but you don't want anyone to know how it works. After all, your invention laid out in detail could inspire someone to get to the same result using an alternative route. You can file to have this invention be a trade secret

You cannot have both a patent and a registered trade secret for an invention. It's one or the other. Both patents and trade secrets provide legal protection, but against different things.

Patents are protected from infringement.

Trade secrets are protected against industrial espionage, e.g. someone sneaking in to your office or hacking your computer and swiping your method.

In a patent, once you've filed and the patent office puts your application up on their website, it's public knowledge. There's no need for someone to dangle from the ceiling in a black catsuit or sit in front of a screen hitting keys randomly to cause a file to download to find out your method. It's public.

If the patent is later granted, infringement protection extends back to the date of filing.

If, on the other hand, you have a trade secret and someone else comes up with the same idea, you have no case against them. Secrets can be developed in parallel without infringing.

But if someone did pull a heist film plot on your registered trade secret, the government will know that it was yours, and if the theft can be proven, then the thief's brand of glow-in-the-dark mouthwash can't be put on the market the way it could be if they had stumbled on the same method you used without ripping off your brand of glow in the dark mouthwash.

That's basically what I know about trade secrets.

If you're trying to figure out whether to apply for patent or for a trade secret, the main question is what matters to you more, holding on to the rights to manufacture or holding on to the knowledge of the process by which you manufacture.

In either case you should talk to an IP lawyer.

Trademarks

Copyrights, patents and trade secrets are each in their own way concerned with the substance of what you have created. Trademarks are about the ability for audiences to recognize that you are the one who made it. Trademarks are names and symbols which identify what is being sold and who is selling it.

Trademarks are meant to be unique identifiers for a particular kind of product. They need to be visually distinct. Ideally, they should also be visually pleasing to buyers of the products. This is why trademarks are usually designed by visual artists specializing in the field.

Trademarks exist to prevent confusion (accidental or deliberate) about which maker's products one is buying. Trademarks must be distinct but only for particular kinds of products. A trademark for furniture and a trademark for frozen foods can look similar because few people make beanbag chairs out of frozen string beans.

Trademark registration and infringement are relatively simple except for trademarking characters. The trademark office uses the same requirements of distinctiveness in appearance for a trademarked character that they do for a logo. You can't trademark a generic dog, but the company that owns {insert famous cartoon dog} can trademark {appearance of that famous cartoon dog}.

If your work uses such characters, you should probably consult a lawyer. Otherwise trademark is a relatively painless process to go through.

Trademarks can be searched, designed, and applied for. The costs are much lower than patents, basically in the same range as copyright registration. Having them can be helpful.

To reiterate, I am not a lawyer. The law can and does change. It's a good idea for every artist to stay as current on the relevant areas of intellectual property law as a lay person can.

Remember that these laws vary from country to country. Don't assume that what one nation does will be done by all. If you know and can consult an IP lawyer on your concerns, that's good, although it can cost lawyer money.

Be aware this chapter is not comprehensive. There are whole law libraries of law and case law on all of these and other kinds of IP. It's complicated out there.

Personal Exercises

1. Go to the websites for your nation's intellectual property registration departments and read their basic documents. In the United States these are the US Patent and Trademark office and the US Copyright Office. Do the same for any other nation's IP law you are interested in.

24

Art as Life

Mountains Are Mountains and Rivers Are Rivers, but They Are Not the Same Mountains and Rivers*

*Variation on the end of a Zen comment that a friend of mine coined years ago.

So where are we, having passed through this discussion of theory and praxes on programming as art?

We're still living in the same real world, still dealing with the same practicalities we were when we began. What, if anything, might be different?

The answers to this are, as with everything in art, both uniquely personal and recurrent through humanity, and as with everything in STEM, a matter of measurement, consideration and application.

Here we hopefully stand and move with those considerations unified as two aspects of one way of looking or one unified perspective with the advantage of double visioned parallax but not paradox.

Looking at the World

Take two steps back. Examine the measurable reality around you. Look, hear, feel and smell the too-much-information.

Feel the toolkits of measurement, the ways that data can be extracted and arranged.

Through Different Artistic Augmentations

Take many one steps forward, in parallel or in series, one invocation after another or many gathered diverse displays of your thinking.

Play with the possible persistences.

What can be made two-back-and-one-forward from where you started?

Where did the different toolkits come from?

DOI: 10.1201/9781003459866-27

Credit Where It's Due. The Fractal Character of Art and Humanity

Look at the tools you use, physical and mental, the products of art and human awareness, measurement, science, engineering and the quiet space-filling music of mathematics.

They were made by people standing in their variations of this perspective who stood back enough to be detached, but close enough to take action.

The histories of art and science are available to be worked from, from moments to events to space time and possibility turning and rolling.

Remember that there were always whos who made these, shared them, corrected and expanded them. Remember there are mortals to be remembered.

Evolving One's Art

Where will you go? What use if any will you make of this?

Practicality and poetry, perhaps to your taste, perhaps not.

There will always be something of use and enjoyability somewhere in the possibilities.

You may walk from here with any possible human reaction. That's the way of the audience.

You may gather utility or error from what I'm saying here. That's the way of the artist.

You may be relieved to be finishing this. I certainly am, but that is a different relief. Art breathed in and out is different from art grown and then released.

Each of us walks away into a different evolution of art and practicality.

The Unplannable Future

We cannot predict what mind will make with what it has been given. So long as mind lasts it will make somethings. It will harvest data from the overflowing possibilities of entangled information and make some use and joy of it for itself and others.

We Never Know but We Can Understand

What we are capable of.

Index

Pages in *italics* refer to figures.

A

acclimatization, 17, 210
Adams, Douglas, 46
aerobic glycolysis, 51
aesthetic variation, 152
algorithmic language (ALGOL), 91–92
animation, 211–212
animators, 121, 212
annoyance, 46–47, 49, 221–222
anthropomorphization, 153
anti-vaxx movement, 237
API programming, 76, 192, 209, 238
APL, 87, 91
apprenticeship, 246, 271
art
    aesthetic theories
        golden ratio, *20*
        John Keats: *Ode on a Grecian Urn*, 19
    artist and audience, 20
        chaotic relationship, 24–26, *25*
    childhood, 16
    composition, 15
        blending, 11
        metaphor, 12
        persistence, 12
        shared properties, 11
        structuring, 12
    contrast, 15
        distinction in properties, 14
        perceivable difference, 12
        separation in environment, 13
        separation in meaning, 13
        separation in space, 12–13
        separation in status, 13
        separation in time, 13
    European, 23
    fashions
        clothing and painting, 17, *18*
        in cooking, 18
        cycles, 17
    history
        images and sculptures, 4
        musical instrument, 3
        music and dance, 4
    as illuminating abstraction, 4–5
        photograph and art of elephant, *5–6*
    interface/surface as expression of, 161
    making personal, 26
    normalization, 16–18
    of persistence, 36
    personal exercises, 26–27
    on practical, personal and cultural levels,
        14–16
    presences in, 122–123
    reality augmented by
        false positives, 10
        human minds, 9, 20
    responsibilities and playspaces
        of artist, 20–22
        of audience, 23–24
    and science, 3, 9, 126, 280
    testable exercises, 27
art as life
    evolving one's art, 280
    fractal character of art and humanity, 280
    looking at the world, 279
    theory and praxes on programming, 279
    through different artistic augmentations, 279
    unplannable future, 280
art–audience interaction, 25
artifacts, 4, 13, 18, 23
artist(s)
    and audience, 20
    cliché of, 239
    conversations between, 158
    copyright to, 204
    cultural pressure, 231
    employing and commissioning other,
        270–271
    human needs of, 270
    ignorance and knowledge of, 21
    organizations
        economic structures, 271
        exploitative structures, 272
        guilds and unions, 272